Oklahoma Notes

Basic Sciences Review for Medical Licensure
Developed at
The University of Oklahoma College of Medicine

Suitable Reviews for:
United States Medical Licensing Examination
(USMLE), Step 1

Oklahoma Notes

Anatomy

Embryology   Neuroanatomy
Gross Anatomy   Microanatomy

Edited by
Raymond E. Papka

With Contributions by
James J. Tomasek   Daniel L. McNeill
Hubert W. Burden   Raymond E. Papka

Springer-Verlag
New York Berlin Heidelberg London Paris
Tokyo Hong Kong Barcelona Budapest

Raymond E. Papka, Ph.D.
Department of Anatomical Sciences
University of Oklahoma
Health Science Center
P.O. Box 26901
Oklahoma City, OK 73190
USA

All contributors share the address above except:
Hubert W. Burden
Department of Anatomy and Cell Biology
East Carolina University
Greenville, NC 27858
USA

Library of Congress Cataloging-in-Publication Data
Anatomy : embryology, gross anatomy, neuroanatomy, microanatomy /
  edited by Raymond E. Papka ; contributors, Raymond E. Papka . . . [et
  al.].
       p.   cm. — (Oklahoma notes)
  Includes bibliographical references.
  ISBN 0-387-94395-1
  1. Human anatomy—Outlines, syllabi, etc.   2. Human anatomy—
Examinations, questions, etc.   3. Embryology, Human—Outlines,
syllabi, etc.   4. Embryology, Human—Examinations, questions, etc.
I. Papka, Raymond E.   II. Series.
  [DNLM: 1. Anatomy—outlines.   2. Anatomy—examination questions.
QS 18 A536 1995]
QM31.A53   1995
611—dc20
DNLM/DLC
for Library of Congress   94-45449

Printed on acid-free paper.

Production managed by Jim Harbison; manufacturing supervised by Jacqui Ashri.
Camera-ready copy prepared by the editor.
Printed and bound by Edwards Brothers, Inc., Ann Arbor, MI.
Printed in the United States of America.

9 8 7 6 5 4 3 2 1

ISBN 0-387-94395-1 Springer-Verlag New York Berlin Heidelberg

Preface to the
Oklahoma Notes

In 1973, the University of Oklahoma College of Medicine instituted a requirement for passage of the Part 1 National Boards for promotion to the third year. To assist students in preparation for this examination, a two-week review of the basic sciences was added to the curriculum in 1975. Ten review texts were written by the faculty: four in anatomical sciences and one each in the other six basic sciences. Self-instructional quizzes were also developed by each discipline and administered during the review period.

The first year the course was instituted the Total Score performance on National Boards Part I increased 60 points, with the relative standing of the school changing from 56th to 9th in the nation. The performance of the class since then has remained near the national candidate mean. This improvement in our own students' performance has been documented (Hyde et al: Performance on NBME Part I examination in relation to policies regarding use of test. J. Med. Educ. 60: 439–443, 1985).

A questionnaire was administered to one of the classes after they had completed the Boards; 82% rated the review books as the most beneficial part of the course. These texts were subsequently rewritten and made available for use by all students of medicine who were preparing for comprehensive examinations in the Basic Medical Sciences. Since their introduction in 1987, over 300,000 copies have been sold. Obviously these texts have proven to be of value. The main reason is that they present a *concise overview* of each discipline, emphasizing the content and concepts most appropriate to the task at hand, i.e., passage of a comprehensive examination over the Basic Medical Sciences.

The recent changes in the licensure examination that have been made to create a Step 1/Step 2/Step 3 process have necessitiated a complete revision of the Oklahoma Notes. This task was begun in the summer of 1991 and has been on-going over the past 3 years. The book you are now holding is a product of that revision. Besides bringing each book up to date, the authors have made every effort to make the tests and review questions conform to the new format of the National Board of Medical Examiners. Thus we have added numerous clinical vignettes and extended match questions. A major revision in the review of the Anatomical Sciences has also been introduced. We have distilled the previous editions' content to the details the authors believe to be of greatest importance and have combined the four texts into a single volume. In addition a book over neurosciences has been added to reflect the emphasis this interdisciplinary field is now receiving.

I hope you will find these review books valuable in your preparation for the licensure exams. Good Luck!

Richard M. Hyde, Ph.D.
Executive Editor

Preface

This book represents a major revision of the *Oklahoma Notes* that were previously produced as individual volumes for the review of Embryology, Gross Anatomy, Neuroanatomy, and Microanatomy. Within this single book, information regarding the four major topics of anatomical sciences has been reorganized and updated. The authors have endeavored to produce a concise and relevant study-book that will provide students a reasonably thorough review of the components of the anatomical sciences. The book is written in outline form in order to reduce the volume of descriptive prose and to facilitate the student's opportunity to get to the "meat" of the subjects.

This review narrative is written expressly for students preparing to take Step 1 of the USMLE. It is a distillation of headlines of essential content for medical Embryology, Gross Anatomy, Neuroanatomy, and Microanatomy. It is written with the assumption that you previously have completed medical school courses that included dissections and microscopy. For most of us, anatomy is quickly forgotten if not used. This outline volume attempts to organize the contents of the anatomy courses into formats useful for review/recall purposes. This is not a traditional review text—there are few illustrations or diagrams. The student is encouraged to thoughtfully engage the narrative, highlight words/phrases as appropriate, and incorporate mental images of structures and their functions as you read. If the mental images fail to focus clearly as you read the narrative, students should consult atlases and more comprehensive texts used in your medical school courses.

Questions for self-examination are placed separately at the end of the book. These questions are intended to be integrative so students should not expect to review a section of the anatomical sciences and then take a self-examination on that section alone.

We wish you the best in your review!

R.E.P.
J.J.T.
D.L.M.
H.W.B.

Contents

Microanatomy   *Raymond E. Papka*

EMBRYOLOGY

GAMETOGENESIS, REPRODUCTIVE CYCLE AND FERTILIZATION

I. CELLS AND CHROMOSOMES

    A. Diploid Cells
    In humans, the cells comprising all renewing cell populations (including the precursors of germ cells) are diploid (2N) and possess a total of 46 chromosomes.

$$
\begin{aligned}
\text{diploid (2N)} &= 46 \text{ chromosomes} \\
\text{autosomal chromosomes} &= 44 \text{ (22 homologous pairs)} \\
\text{sex chromosomes} &= 2 \\
\text{females} &= \text{XX (homologous)} \\
\text{males} &= \text{XY (nonhomologous)}
\end{aligned}
$$

    B. Haploid Cells (Gametes)
    The gametes (sperm and ova) are highly specialized reproductive cells containing only one-half (N or 23) the number of chromosomes found in renewing cell populations. The reduction in chromosome number is accomplished by two specialized cell cycles which are referred to as Meiosis I and Meiosis II.

$$
\begin{aligned}
\text{haploid (N)} &= 23 \text{ chromosomes} \\
\text{autosomal chromosomes} &= 22 \text{ (one member from each pair)} \\
\text{sex chromosomes} &= 1 \\
\text{females} &= \text{X (always)} \\
\text{males} &= \text{X or Y}
\end{aligned}
$$

    C. Chromosome Morphology
        1. centromeres
        2. arms
        3. chromatids

    D. Mitosis
    The two most important events occurring during **mitosis** are DNA replication (interphase) and centromeric division (late metaphase). When DNA replication and centromeric division occur during the **same** cell cycle, the number of chromosomes in both daughter cells will remain constant (**diploid**) and the daughter cells will be identical genetically.

    E. Meiosis
    During the two specialized cell cycles of gametogenesis (meiosis I and meiosis II), DNA replication and centromeric division occur in separate cell cycles. DNA replication occurs during meiosis I; centromeric division occurs during meiosis II. When cells divide without centromeric division, the number of chromosomes transmitted to daughter cells must be reduced by one-half. The resulting haploid daughter cells will **NOT** be genetically identical, each cell receives only one member (maternal or paternal) from each of the 23 chromosome pairs.

II. ABNORMALITIES IN CHROMOSOME NUMBER (ANEUPLOID CELLS)

A. Nondisjunction

The most common event producing abnormal chromosome numbers (aneuploidy) is **nondisjunction**. When nondisjunction occurs, the chromosomes are unequally distributed between the daughter cells so that one cell receives an extra chromosome (trisomy) and the other cell is missing that chromosome (monosomy). **NONDISJUNCTION CAN OCCUR ANYTIME A CELL DIVIDES (MITOTICALLY OR MEIOTICALLY) AND CAN AFFECT EITHER AUTOSOMAL OR SEX CHROMOSOMES.** It is generally accepted that one of the most frequent causes for nondisjunction is delayed or asynchronous division of centromeres during late metaphase.

B. Meiotic Nondisjunction

During gametogenesis, nondisjunction results in the production of two abnormal gametes (haploid plus one or haploid minus one chromosome). If one of the abnormal gametes participates in fertilization an aneuploid zygote is produced which either possesses an extra chromosome (sex or autosomal trisomy) or lacks a chromosome (sex or autosomal monosomy). Individuals developing from these abnormal zygotes will be either trisomic or monosomic for the chromosome involved.

1. Zygotes with autosomal aneuploidy. Although 22 pairs of autosomal chromosomes are present in human zygotes and any chromosome can be affected by nondisjunction (all have been reported in aborted material), only a few (trisomic only) complete development to present as postnatal medical problems.

   a) Autosomal Monosomy. True monosomic conditions involving any of the 22 autosomal chromosomes have not been reported to complete intrauterine development and are generally considered to be invariably lethal.

   b) Autosomal Trisomy. Trisomic conditions for autosomal chromosomes are usually but not invariably lethal. The only common autosomal trisomic condition with long term survival is trisomy 21 (Down's syndrome). Trisomy 13 (Patau's syndrome) and trisomy 18 (Edward's syndrome) are occasionally reported but these individuals rarely survive beyond the first few months of postnatal life.

   c) Maternal Age. The increased frequency of nondisjunction with advancing maternal age is usually attributed to the fact that the first meiotic division of all ova begins during fetal life. Ova produced during the later years of the reproductive life span are very old and have been subjected to the cumulative environmental (radiation, chemicals, therapeutic drugs, viral infections) and physiological (endocrine changes) effects of aging.

2. Zygotes with Sex Chromosome Aneuploidy. In contrast to the small number of aneuploid conditions resulting from unequal distribution of the 22 pairs of autosomes, nondisjunction affecting the single pair of sex chromosomes may produce zygotes and adult individuals of at least four basic genotypes. Monosomic and trisomic conditions for sex chromosomes are well known clinically and are relatively common.

SEX CHROMOSOME NONDISJUNCTION

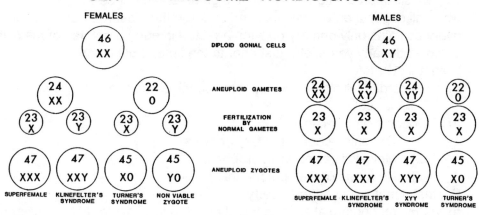

C. Mitotic Nondisjunction

Mitotic nondisjunction during the earliest stages of embryonic development may result in **aneuploid individuals with mixed populations of somatic cells**, i.e., mixed karyotypes; individuals originating from mitotic nondisjunction are referred to as **mosaics**. When mitotic nondisjunction occurs in what was initially a normal diploid embryo, the body of the surviving mosaic individual is comprised of a mixed population of normal and abnormal (aneuploid) cells. The presence of normal cell populations in mosaics increases the chances of survival by reducing the severity of the coexisting aneuploidy, e.g., low fertility rather than complete sterility.

III. GAMETOGENESIS

A. Primordial Germ Cells

Primordial germ cells are the progenitor cells of the male and female gametes. They arise from allantoic endoderm and migrate to the developing gonad. In males they form spermatogonia. In females they form oogonia.

B. Spermatogenesis

1. Spermatogenesis is the production of the male gamete, the sperm. Spermatogenesis begins at puberty with the rise in testosterone and continues into advanced age.
2. Spermatogonia are transformed into spermatozoa.
   a) Occurs after puberty.
   b) Spermatogonia enlarge and transform into primary spermatocytes.
   c) First meiotic division produces two secondary spermatocytes.
   d) Second meiotic division produces four spermatids.
   e) Spermatogenesis takes about 2 months to complete.
   f) Sertoli cells. Spermatogenesis occurs in microenvironment provided by Sertoli cells.
3. Spermatogonial stem cells. Persist throughout life to continuously produce spermatogonia.
4. Spermiogenesis. Differentiation of spermatid into mature sperm.

C. Oogenesis
   1. Oogenesis is production of the female gamete. Oogenesis begins during late fetal life when all oogonia enter meiosis I and undergo their last replication of DNA. At birth the ovary contains only primary oocytes arrested in early prophase of meiosis I; meiosis I is completed many years later just prior to the time of ovulation.
   2. Prenatal events.
      a) Primordial follicles form. Oogonia become surrounded by a single layer of follicular cells.
      b) Oogonia enter meiosis I.
         (1) Oogonia become primary oocytes
         (2) Meiosis is arrested in prophase I.
         (3) No further development until puberty.
   3. Postnatal maturation after puberty
      a) Several primordial follicles are stimulated to begin to mature every month at the beginning of each reproductive cycle.
      b) Follicular cells thicken and grow to form primary follicle
      c) Primary oocyte becomes surrounded by zona pellucida.
      d) Maturation of primary follicle.
         (1) Growing follicle
         (2) Antral follicle
      e) Maturation of primary oocyte:
         (1) completes first meiotic division just prior to ovulation.
         (2) forms secondary oocyte and first polar body.
         (3) secondary oocyte enters second meiotic division.
         (4) arrested at metaphase II.
   4. Ovulation
      a) Release of secondary oocyte with zona pellucida and corona radiata.
      b) Corpus luteum. Remaining "ovulated" follicle
   5. Fertilization. Fertilization is necessary for the secondary oocyte to complete the second meiotic division and become an oocyte.

IV. FEMALE REPRODUCTIVE CYCLE

   A. Reproductive Cycle
   Human females undergo sexual cycles which prepare the reproductive system for pregnancy; these cycles are controlled by the combined action of the hypothalamus, pituitary, ovary and uterus.
      1. Ovarian cycle. Regulated as a direct response to pituitary gonadotropins.
         a) follicle stimulating hormone (FSH)
         b) luteinizing hormone (LH)
      2. Uterine cycle. Regulated as a direct response to hormones produced by the ovary.
         a) estrogen
         b) progesterone

B. Ovarian Cycle
   1. Initial phase of follicular development. This occurs in response to relatively low and tonic secretion of FSH and LH by the pituitary gland. This phase culminates in increasing blood levels of estrogen produced by the maturing follicle.
   2. A midcycle surge of gonadotropins. This phase is elicited by the rising blood levels of estrogen. The surge of gonadotropins, particularly LH, results in ovulation.
   3. A post-ovulatory phase of high circulating progesterone levels and low circulating gonadotropin levels. Progesterone is secreted by the corpus luteum. LH promotes steroidogenesis of the corpus luteum.

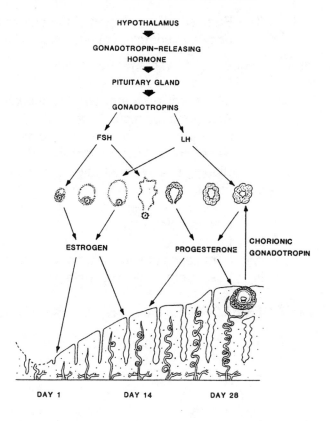

C. Uterine Cycle
   1. Proliferative phase. This phase is under control of estrogen produced by follicle.
   2. Secretory phase. This phase is under control of progesterone secreted by corpus luteum.
   3. Menstruation. This phase occurs as a result of decreased levels of progesterone and estrogen.

D. Pregnancy
   The reproductive cycle is interrupted by pregnancy.
   1. Human chorionic gonadotropin. This LH-like hormone is produced by the trophoblastic cells of the conceptus.
   2. Corpus luteum. Chorionic gonadotropin rescues the corpus luteum from impending regression and maintains its continued production of progesterone during early pregnancy. A functional corpus luteum is essential for early gestation, until the placenta can assume the endocrine activities at about 6 weeks of development.

V. FERTILIZATION

   A. Three Events Necessary for Successful Fertilization
      1. Sperm/oocyte recognition. The sperm must recognize the secondary oocyte and fuse only with it.
      2. Prevention of polyspermy. Only one sperm must enter the oocyte otherwise polyploidy will result.

3. Egg activation. Fusion of the sperm with the oocyte must activate the secondary oocyte to complete meiosis II and begin development.

B. Sequence of Events for Fertilization
1. Capacitation of spermatozoa. This occurs in the female reproductive tract and is necessary for sperm/oocyte recognition.
2. Acrosome reaction. Proteolytic enzymes are released that are needed to penetrate the corona radiata and zona pellucida.
3. Penetration of corona radiata.
4. Binding to and penetration of zona pellucida.
5. Fusion of sperm cell and secondary oocyte cell membranes.
6. Cortical granule reaction. Release of cortical granules alters the zona pellucida and prevents polyspermy.
7. Egg activation.
8. Stimulation of secondary oocyte to complete meiosis II.
9. Fusion of female and male pronuclei. Formation of diploid zygote.

EARLY DEVELOPMENT OF THE CONCEPTUS: WEEKS 1-2

I. WEEK 1

A. Cleavage
1. The zygote is subdivided by a series of **rapid mitotic divisions** referred to as cleavage.
2. The 1st cleavage appears 24 hrs after fertilization, with others occurring about every 20 hrs.
3. The new cells formed are **blastomeres**.
4. The mitosis is unusual in that division is not accompanied by growth. The cytoplasm is partitioned into smaller blastomeres.
5. Cleavage occurs within the uterine tube.
6. By approximately the 3rd day the conceptus is referred to as a **morula**. At this time it enters the uterine cavity.

B. Compaction
1. The peripheral blastomeres maximize adhesion and form tight junctions.
2. Compaction segregates blastomeres into two populations:
   a) **outer cell mass**.
   b) **inner cell mass**.
3. Compaction occurs around 4 days after fertilization at approximately the 16 cell stage.

C. Blastocyst Formation
1. Blastocyst forms approximately 5 days after fertilization.
2. **Blastocyst cavity** forms, a hollow, fluid filled cavity.
3. Two populations of cells are formed.
   a) Outer cell mass forms outer epithelial layer, the **trophoblast**, the primary source for the placenta.
   b) Inner cell mass forms inner group of cells, the **embryoblast**, gives rise to the embryo proper.

D. Implantation
1. Blastocyst **hatches** from zona pellucida approximately 6 days after fertilization.
2. Trophoblast contacts endometrial wall.
3. Blastocyst invades into endometrial wall.
4. Trophoblast forms two cell types.
    a) **Cytotrophoblast**
        (1) Actively dividing cells.
        (2) Gives rise to syncytiotrophoblast.
    b) **Syncytiotrophoblast**
        (1) Large syncytium.
        (2) Highly invasive. The syncytiotrophoblast is responsible for invasion of the blastocyst into the endometrial wall.
        (3) Syncytiotrophoblast is an endocrine organ. Produces protein and steroid hormones.
            (a) Produces large amounts of **human chorion gonadotropin** (hCG).
            (b) hCG is responsible for maintenance of the corpus luteum.

II. IMPLANTATION SITES

A. Intrauterine Implantation Sites
1. The blastocyst normally implants in the superior part of body of uterus and most frequently in the posterior wall.
2. Other implantation sites in the uterus are referred to as abnormal. Placenta previa occurs when the blastocyst implants in the inferior part of body of uterus and placenta develops covering the internal os of the cervix.

B. Extrauterine Implantation Sites
1. **Ectopic pregnancy**. Implantation of the blastocyst outside the uterine cavity.
2. **Tubal implantation**.
    a) Most common ectopic pregnancy is tubal implantation.
    b) Tubal implantation is the result of delayed transport of zygote. This may be result of chronic inflammation of the uterine tube with partial destruction of tubal mucosa (endometriosis, gonorrhea, tuberculosis).
3. Ovarian and abdominal implantation are very rare.

III. WEEK 2

A. Formation of the Bilaminar Embryo
1. Embryoblast rearranges to form a flattened disk consisting of two layers of simple epithelium arranged basal-to-basal surface.
2. **Epiblast**. The upper layer of the bilaminar embryo.
    a) Forms the floor of the amniotic cavity.
    b) Epiblast is continuous at periphery of embryonic disk with amniotic epithelium.
3. **Hypoblast**. The lower layer of the bilaminar embryo.
    a) Forms the roof of primary yolk sac.
    b) Hypoblast is continuous at periphery of embryonic disk with exocoelomic membrane.

B. Formation of the Amniotic Cavity
1. Amniotic cavity forms as a fluid filled cavity between epiblast cells.

2. Amniocytes. Epiblast cells forming roof of amniotic cavity.

C. Formation of the Yolk Sac and Extraembryonic Coelom
1. **Primary yolk sac**
a) Hypoblast migrates and lines blastocyst cavity to form primary yolk sac.
b) Exocoelomic membrane forms lining of primary yolk sac.
2. **Extraembryonic mesoderm**
a) Appears earlier and originates from a source different than that of the embryonic mesoderm.
b) Fills space between cytotrophoblast and exocoelomic membrane.
3. **Extraembryonic Coelom (Chorionic cavity)**
a) extraembryonic mesoderm splits to form new cavity
b) extraembryonic coelom lined by extraembryonic mesoderm
c) connecting stalk suspends embryonic disk, amnion and yolk sac in extraembryonic coelom
4. **Secondary (definitive) yolk sac**
a) A new wave of hypoblast migrates and lines primary yolk sac to form secondary yolk sac.
b) Secondary extraembryonic endoderm. The lining of secondary yolk sac.
c) Primary yolk sac degenerates.
5. **Chorion**
a) Forms the outermost layer of conceptus.
b) Formed by syncytiotrophoblast, cytotrophoblast and extraembryonic mesoderm.

D. Primitive Uteroplacental Circulation
1. **Trophoblastic lacunae.** Spaces form and interconnect within syncytiotrophoblast. Syncytiotrophoblast invades maternal vessels. Maternal vessels connect with lacunae which fill with maternal blood.
2. Provide nutrient and waste exchange with maternal circulation.
3. Later in develop trophoblastic lacunae will coalesce to form the intervillous vascular channels of the placenta.

EXTRAEMBRYONIC MEMBRANES AND UMBILICAL CORD

I. EXTRAEMBRYONIC MEMBRANES
The extraembryonic membranes are temporary structures necessary for embryonic development; they are lost as part of the afterbirth at the time of parturition. They appear and become functional (partially at least) during the second week of development.

A. Chorion
1. Forms the outermost extraembryonic membrane.
2. Comprised of three layers:
a) **syncytiotrophoblast**.
b) **cytotrophoblast**.
c) **extraembryonic mesoderm**.
3. Later a portion of the chorion will form the **fetal portion of the placenta**.

B. Amnion
1. Forms the layer surrounding the amniotic cavity.

2. Comprised of two layers:
   a) **amniocytes** derived from epiblast.
   b) extraembryonic mesoderm.
3. Amniotic cavity
   a) Expands to surround developing embryo.
   b) Obliterates extraembryonic coelom.
   c) Amniotic membrane and chorion fuse to form amniochorionic membrane.
   d) Amniotic membrane forms covering of umbilical cord.
4. Amniotic fluid
   a) Fetus contributes to amniotic fluid by excreting approximately 500 ml of urine daily.
   b) Subsequent swallowing of amniotic fluid by the fetus followed by intestinal absorption and elimination via the placenta prevents excessive accumulation.
5. **Oligohydramnios**.
   a) Abnormally small amounts of amniotic fluid.
   b) May be result of decreased fetal secretion of urine. May be the result of **urethral obstruction** or **renal agenesis**.
6. **Polyhydramnios**.
   a) Abnormally large amounts of amniotic fluid.
   b) May be result of decreased fetal swallowing of amniotic fluid. May be the result of **high intestinal obstruction, i.e. esophageal, pyloric, or duodenal atresia or stenosis, or anencephaly**.
7. Significance of amniotic fluid.
   a) Permits symmetrical external growth of the fetus.
   b) Enables the fetus to move freely, important in muscular development.
   c) Prevents adherence of amnion to embryo and fetus.
   d) Plays a role in normal fetal lung development.
8. **Diagnostic amniocentesis**.
   a) Performed approximately 14 weeks after last normal menstrual period.
   b) Common technique for **detecting genetic disorders**.
   c) **Alpha-fetoprotein assay**. High levels suggestive of neural tube defects or ventral wall defects.

C. Yolk Sac
   1. The yolk sac is essentially nonfunctional.
   2. Important in early development.
      a) Source of **primordial germ cells**.
      b) Role in hematopoietic functions.
      c) Involved in gut tube formation.
   3. Becomes located in umbilical cord and degenerates.
   4. Yolk stalk (vitelline duct).
      a) Connects the gut tube to the yolk sac.
      b) Normally degenerates.
      c) Failure of the yolk stalk to degenerate results in congenital malformations.
         (1) Complete failure may result in an **ileo-umbilical fistula** or a **fibrous cord**.
         (2) Partial failure results in **Meckel's diverticulum**.

D. Allantois
   1. Develops as a diverticulum off secondary yolk sac that projects into body stalk.

2. The allantois small and rudimentary in humans.
3. Becomes incorporated into ventral body and forms the **median umbilical ligament**.
4. Failure of the allantois to degenerate results in congenital malformations, i.e. **urachal fistula or urachal cysts**.

II. UMBILICAL CORD

    A. The umbilical cord is a composite structure formed by contributions from the:
        1. connecting stalk (body stalk) which contains the allantois and the umbilical arteries and vein.
        2. yolk stalk and sac.
        3. portions of the amnion ensheathing all of the above structures.

    B. Umbilical vessels
        1. Two umbilical arteries.
        2. The left umbilical vein persists.
        3. The right umbilical vein degenerates.

    C. Size
        1. The normal length ranges from 50-55 cm. Exceptionally long umbilical cords are predisposed to forming knots.
        2. The normal diameter ranges from 1-2 cm.

PLACENTA

I. FORMATION

    A. Maternal Contributions - **Decidua**
        1. **Decidual reaction**. Implantation stimulates changes in uterine endometrium resulting in the accumulation of lipids and glycogen in endometrial stomal cells.
        2. The uterine mucosa is referred to as the decidua and is divided into three areas determined by their relationship to the implanted blastocyst.
        3. **Decidua basalis**.
            a) The decidua located below or deep to implanted blastocyst.
            b) The decidua in this area forms the **maternal portion of the placenta**
        4. **Decidua capsularis**.
            a) Decidua covering or superficial to implanted blastocyst.
            b) Subsequently fuses with decidua parietalis to obliterate uterine cavity.
        5. **Decidua parietalis**. Decidua other than site of implantation.
        6. Decidua is shed as part of afterbirth during parturition.

    B. Fetal Contributions - **Chorion**
        1. **Trophoblastic lacunae**.
            a) Form with syncytiotrophoblast during the second week of development.
            b) Trophoblastic lacunae within syncytiotrophoblast become filled with maternal blood.
            c) These coalesce to form the **intervillous spaces of the placenta**.
        2. **Chorionic villi**.
            a) Extend and grow into trophoblastic lacunae to bring embryonic and maternal blood in close proximity.
            b) **Primary chorionic villi** - composed of cytotrophoblast and syncytiotrophoblast.

        c)   **Secondary chorionic villi** - core of extraembryonic mesoderm.

        d)   **Tertiary chorionic villi** - vascularization of mesodermal core which appears at end of third week.

    3.   **Chorion laeve** (smooth chorion) - Chorionic villi on chorion associated with decidua capsularis are lost and the chorion becomes smooth.

    4.   **Chorion frondosum**.

        a)   Chorionic villi on chorion associated with decidua basalis persist to form anchoring and branch villi of the placenta.

        b)   Forms the **fetal portion of placenta**.

C.   Mature Placenta

    1.   The placenta is fully formed and mature by 4th month of development.

    2.   Chorionic plate.

        a)   Composed of layers of the chorion.

        b)   Placental villi project into intervillous spaces filled with maternal blood.

    3.   Size.

        a)   Diameter of 15-20 cm.

        b)   Total volume of approximately 500 ml.

        c)   Maternal blood volume of approximately 150 ml.

        d)   Maternal blood flow of approximately 600 ml/min.

        e)   Weight of 500-600 grams (approximately one-sixth of the fetal weight).

II.   PLACENTAL CIRCULATION

A.   Fetal Blood

    1.   **Deoxygenated fetal blood**. Leaves the fetus via the umbilical arteries and passes into the capillaries in the chorionic villi where gaseous and nutrient exchange occurs.

    2.   **Oxygenated fetal blood**. Returns to the fetus via the umbilical vein.

B.   Maternal Blood

    1.   Fills the intervillous spaces.

    2.   Volume is about 150 ml of maternal blood and is replaced about three to four times per minute.

III.   PLACENTAL MEMBRANE

A.   Function. Separates maternal and fetal blood. Exchange of respiratory gases, waste products, and antibodies between fetal and maternal blood occurs through the placental membrane.

B.   Consists of 4 Layers.

    1.   Syncytiotrophoblast.

    2.   Cytotrophoblast (Langhan's layer).

    3.   Connective tissue of the villus core (extraembryonic somatic mesoderm).

    4.   Endothelium of the fetal capillaries.

C.   Thinning of Placental Membrane

    1.   Occurs after 4th month.

    2.   Increases efficiency of exchange between fetal and maternal blood.

    3.   Thinning is the result of disappearance of cytotrophoblast layer and thinning of connective tissue.

IV. PLACENTAL FUNCTIONS

    A. Metabolic Transport
        1. Respiratory gases. O_2 and CO_2 cross by simple diffusion.
        2. Nutrients.
        3. Waste products.

    B. Antibodies
        1. Newborns have **passive humoral immunity** acquired from the mother by selective transfer of IgGs. IgA and IgM antibodies do not cross the placental barrier.
        2. **Hemolytic disease of the newborn (erythroblastosis fetalis)**
            a) Occurs when the mother is Rh-negative and the fetus is Rh-positive.
            b) If the mother becomes sensitized to the Rh positive fetal cells, she responds by producing anti-Rh-antibodies of the IgG type which are then transferred across the placental barrier with subsequent hemolysis of the fetal red blood cells.

    C. Endocrine Functions
        1. Protein hormones. Chorion gonadotropin produced by the syncytiotrophoblast appears in maternal urine during the second week (during implantation) and is present in large amounts until the fourth month of pregnancy.
        2. Steroid hormones. Progesterone is produced throughout pregnancy from precursors of maternal origin; estrogens are synthesized from precursors supplied by the fetal liver and adrenal cortex.

    D. Pathogens
        1. Various pathogens can cross the placental membrane and infect the fetus in utero
        2. Viruses - rubella, cytomegalovirus, HIV
        3. Bacteria - treponema pallidum (syphilis)
        4. Protozoa - toxoplasma
        5. These may induce abortion or cause wide range of congenital abnormalities.

    E. Teratogens
        1. Teratogens cross the placenta membrane and cause malformations.
        2. They act predominantly during a **critical period** of development. Usually during embryonic period of development (organogenetic period).
        3. Many therapeutic drugs are teratogens, i.e. warfarin, dilantin, retinoids, synthetic steroids, tetracycline, many of the chemotherapeutic agents, thalidomide.
        4. Many social drugs are teratogens.
            a) **Alcohol is believed to be the leading cause for congenital mental retardation.**
            b) Tobacco and cocaine.

FORMATION OF THE TRILAMINAR EMBRYO

I. GASTRULATION

    A. Gastrulation is the process by which the bilaminar embryonic disc is transformed into a three-layered or trilaminar embryonic disc. Three germ layers are formed by the process of gastrulation.
        1. **Ectoderm**.
        2. **Mesoderm**.

3. **Endoderm**.
4. All three germ layers arise from the epiblast.

B. Primitive Streak
   1. Appears about day 14.
   2. Thickened midline band of epiblast in caudal part of bilaminar disc.
   3. Comprised of:
      a) primitive groove - midline groove
      b) primitive pit - depression at cranial end of primitive groove
      c) primitive node (Hensen's node) - mound surrounding primitive pit
   4. Defines primary body axes of embryo.
   5. Ingression of epiblast through primitive groove and pit to form mesoderm and endoderm.
   6. Epithelial epiblast cells transform to migratory mesenchymal cells as they ingress through the primitive streak.
   7. Fate of primitive streak.
      a) Regresses caudally.
      b) Closes in sacrococcygeal region.
      c) Failure to regress may result in sacrococcygeal teratomas.

C. Embryonic Ectoderm Formation
   Epiblast cells that do not ingress through the primitive streak will form the ectodermal germ layer.

D. Embryonic Endoderm Formation
   Early epiblast cells that ingress through the primitive streak and invade and replace the hypoblast of the embryonic disc with a new layer of cells will form the endodermal germ layer.

E. Embryonic Mesoderm Formation
   Epiblast cells that ingress through primitive streak and migrate between epiblast and endoderm will form the mesodermal germ layer.
   1. Mesodermal cells migrate throughout embryonic disc **EXCEPT** in those regions where the ectoderm and endoderm are fused. These areas are:
      a) **Oropharyngeal membrane (prochordal plate).**
      b) **Cloacal membrane**.
   2. Mesodermal cells migrate cranial to oropharyngeal membrane to form the **cardiogenic region**.
   3. Mesodermal cells migrate caudal to cloacal membrane to form the future ventral body wall caudal to the umbilicus.
   4. Mesodermal cells migrate laterally to fuse with extraembryonic mesoderm

II. ORGANIZATION OF EMBRYONIC MESODERM

A. Notochord
   1. **Formation**. Epiblast ingresses through the primitive pit and migrates cranially along midline to oropharyngeal membrane to form notochordal process. The notochordal process undergoes a transformation to form the notochord.
   2. **Developmental importance**. The appearance of the notochord is the first indication of the future axial skeleton and defines the primary body axis. The notochord eventually

extends from the oropharyngeal membrane to the cloacal membrane. It induces the formation of the musculoskeletal and nervous systems.

3. **Fate**. The notochord regresses except for the formation of the nucleus pulposus of intervertebral disks.

B. Mesodermal cells migrating laterally from the primitive streak form paraxial, intermediate and lateral plate mesoderm.

C. Paraxial Mesoderm
   1. Location.
      a) Embryonic mesoderm oriented parallel and immediately adjacent to notochord.
      b) Forms paired bands of mesoderm.
      c) Extends from oropharyngeal membrane to primitive streak.
   2. Gives rise to:
      a) **All of axial skeleton (except viscerocranium of skull).**
      b) **All skeletal musculature**.
      c) **Portion of dermis**.
   3. Segmentation of paraxial mesoderm.
      a) Paraxial mesoderm. Initially forms somitomeres followed by segmentally arranged somites.
         (1) **Somitomeres**. Transitory, indistinct groupings of paraxial mesoderm.
         (2) **Somites**. Discrete blocks of segmentally arranged paraxial mesoderm.
      b) **Cephalic paraxial mesoderm**.
         (1) Forms 7 somitomeres.
         (2) Cephalic somitomeres do not develop into somites.
         (3) Give rise to structures in head and neck.
      c) First somites appear in future region of base of skull around day 20.
      d) Other somites develop in cranial to caudal sequence.
      e) 42-44 somites are formed.
      f) **Somites establish segmental organization of body**. 4 occipital somites, 8 cervical somites, 12 thoracic somites, 5 lumbar somites, 5 sacral somites, 3 coccygeal somites, last 5-7 somites regress and disappear.
   4. Somites disperse to form two groups of cells.
      a) **Sclerotome**.
         (1) Skeletal forming portion of somite.
         (2) Each sclerotome will contribute to the formation of the vertebrae, intervertebral disks, ribs and sternum of the axial skeleton.
      b) **Dermamyotome**. Splits into myotome and dermatome.
         (1) **Myotome**.
         (2) Muscle forming portion of somites.
            (a) Forms **all** of the skeletal muscle.
            (b) **Dermatome**. Integumentary (dermis) forming portion of somites.

D. Intermediate Mesoderm
   1. Location
      a) Embryonic mesoderm oriented parallel and just lateral to paraxial mesoderm.
      b) Forms paired bands of mesoderm.
      c) Extends from oropharyngeal membrane to primitive streak.
   2. Gives rise to portions of the urinary and reproductive system.

E.   Lateral Plate Mesoderm
  1.   Location
    a)   Embryonic mesoderm extends laterally from intermediate mesoderm.
    b)   Extends cranial to oropharyngeal membrane and fuses across midline.
    c)   Extends caudal to cloacal membrane and fuses across midline.
  2.   Divides into 2 layers separated by embryonic coelom.
    a)   **Somatic mesoderm**.
      (1)   Associated with ectoderm - somatopleure.
      (2)   Contributes to adult body wall.
      (3)   Forms parietal serous membrane lining the body cavities.
    b)   **Splanchnic mesoderm**.
      (1)   Associated with endoderm - splanchnopleure.
      (2)   Contributes to form wall of gut tube.
      (3)   Forms visceral serous membrane lining the body cavities.
  3.   **Embryonic coelom** is located between two layers of lateral plate mesoderm. The embryonic coelom will form the body cavities - pericardial, pleural and peritoneal cavities.

DEVELOPMENT OF GENERAL BODY FORM, BODY CAVITIES AND DIAPHRAGM

I.   FOLDING OF THE EMBRYO - ESTABLISHMENT OF BODY FORM

  A.   Folding of Trilaminar Embryo
    1.   Body folding is the result of greater **differential growth** in the midline of the trilaminar embryo than that occurring in peripheral areas.
    2.   Body folding transforms flattened trilaminar embryonic disc into cylindrical embryo.
    3.   After body folding embryo exhibits **basic tubular configuration of adult organism**.

  B.   Median Folding - Head and Tail Folds
    1.   Median folding is the result of **craniocaudal elongation** of notochord and neural tube.
    2.   **Head folding** "swings" cranial structures ventrally and caudally. The major structures moved ventrally and caudally by head folding are:
      a)   **Septum transversum** - future diaphragm.
      b)   **Cardiogenic region and future pericardial cavity.**
      c)   **Oropharyngeal membrane**.
    3.   **Tail folding** "swings" cloacal membrane ventrally and cranially.
    4.   Median folding results in the upper part of yolk sac being incorporated into embryo as the primitive gut tube.

  C.   Horizontal Folding - Lateral Folds
    1.   Lateral edges of embryonic disk move ventrally and medially.
    2.   Lateral folding results in the formation of ventral body wall.

  D.   Formation of the Umbilicus
    1.   Lateral edges of embryonic disk are pulled around ventral umbilicus like a purse-string.
    2.   Emerging from umbilicus are yolk sac stalk and connecting stalk.

II.   FORMATION OF PRIMITIVE GUT

  A.   Body folding results in the formation of an endoderm lined gut tube.

B. Foregut
   1. Endodermal blind-ending tube projecting cranially into region formed by head fold.
   2. Digestive (dorsal) portion forms pharynx, esophagus and stomach.
   3. Respiratory (ventral) portion forms trachea and lungs.

C. Midgut
   1. Primitive midgut is not a complete tubular structure.
   2. Opens via vitelline duct (yolk stalk) into yolk sac through umbilical ring.
   3. Forms most of small intestine and part of colon.

D. Hindgut
   1. Endodermal blind-ending tube projecting caudally into region formed by tail fold.
   2. Caudal portion expands to form cloaca.
   3. Allantois projects off cloaca into connecting stalk.
   4. Forms remaining part of colon, rectum and urinary bladder.

III. EMBRYONIC COELOM

A. Formation
   1. Embryonic coelom formation begins in the trilaminar embryo during the third week of development and coincides with late stages of embryonic mesoderm development.
   2. Embryonic coelom is formed by splitting of lateral plate mesoderm splits into two layers.
      a) **Somatic mesoderm** - associated with ectoderm.
      b) **Splanchnic mesoderm** - associated with endoderm.

B. Coelomic Shape
   1. Forms inverted "U" configuration.
      a) Paired right and left coelomic cavities.
      b) Coelomic cavities continuous across midline cranial to oropharyngeal membrane - cardiogenic region.
   2. **Primitive pericardial cavity** - cranial to oropharyngeal membrane where right and left coelomic cavities meet in midline.
   3. **Pericardioperitoneal canals** - paired right and left canals connecting pericardial and peritoneal cavities.
   4. **Peritoneal cavity** - communication of right and left coelomic cavities with extraembryonic coelom.

C. Coelomic Communications
   1. Embryonic coelom extends laterally to become continuous with the extraembryonic coelom in region of future peritoneal cavity.
   2. These communications are important because they will ultimately be located at the umbilicus and allow the herniating midgut loop access to the adjacent umbilical (extraembryonic) coelom.
   3. Communication of extraembryonic and embryonic coelom does not occur in cranial portion of embryonic disk.
   4. Embryonic and extraembryonic coelom separated by septum transversum.

D. Septum Transversum
   1. Most cranial structure in trilaminar embryonic disk.
   2. Defines either the cranial end (before folding) or caudal end (after folding) of pericardial cavity.

3. Participates in formation of the diaphragm.

IV. MOVEMENTS OF COELOM

A. Head Fold. Movement of septum transversum and pericardial cavity ventrally and caudally.

B. Lateral Folds. Formation of ventral mesentery suspending gut tube in peritoneal cavity.

V. PARTITIONING OF COELOMIC SPACE

A. Pericardial Cavity
1. Primitive pericardial cavity is separated into:
    a) **Ventral definitive pericardial cavity**.
    b) **Paired dorsolateral pleural cavities**.
2. **Pleuropericardial membranes** separate primitive pericardial cavity into pericardial cavity and paired pleural cavities.
    a) Pleuropericardial membranes originate from lateral body wall.
    b) They grow medially to fuse with each other and primitive mediastinum.
    c) **Serous parietal pericardium and mediastinal pleura** form from pleuropericardial membranes.
3. Most of the 'definitive pleuropericardial membranes' separating the heart and lungs of the adult are new structures split from the internal surface of the body wall by the developing lungs. It is the lateral and ventral expansion of the lungs and pleural cavities within the body wall that moves the heart and pericardial cavity to its definitive location in the mediastinum and allows the pleural structures to eventually become the most anterior visceral structures in the thorax.

B. Pleural and Peritoneal Cavities
1. The pleural and peritoneal cavities are separated by the formation of right and left **pleuroperitoneal membranes**.
2. Pleuroperitoneal membranes originate from lateral body wall and dorsal edge of the septum transversum and grow dorsomedially to fuse with the splanchnic mesoderm of the dorsal wall of the esophagus (mesoesophagus).
3. Pleuroperitoneal membranes participate in the formation of the diaphragm.

VI. DIAPHRAGM FORMATION

A. The adult diaphragm is a composite structure and forms from:
1. **Septum transversum**.
2. **Pleuroperitoneal membranes**.
3. **Splanchnic mesoderm surrounding esophagus** (primitive mediastinum).
4. **Body wall**. Produced as a result of lung and pleural cavity growth within the somatopleura; these areas receive their sensory innervation via somatopleuric intercostal nerves.

B. Innervation
1. The skeletal muscle for all parts of the diaphragm is derived from cervical myotomes (primarily C-4).
    a) Myoblasts from the cervical region invade the septum transversum.
    b) Head folding and differential growth move the septum transversum caudally to the adult location.
2. Muscle innervation is via the **phrenic nerve** (C3-C5).

C. Congenital Diaphragmatic Hernia
1. Results from a defect in the formation of pleuroperitoneal membrane.
2. The defect is always in the dorsal aspect of diaphragm.
3. The defect usually occurs on left side.
4. The herniation of abdominal viscera through the defect and into the thoracic cavity may compromise fetal lung development and result in **pulmonary hypoplasia**.

NERVOUS SYSTEM

I. NEURAL TUBE FORMATION

A. Neurulation
1. Neural plate formation
a) Axial mesoderm (notochord and paraxial) **induces** the overlying ectoderm to thicken and form the neural plate.
b) The neural plate appears during 3rd week of development.
c) **Neuroectoderm** - the ectodermal cells which comprise the neural plate.
2. Neurulation
a) **Neurulation** - process by which neural plate is converted into the neural tube.
b) The lateral edges of neural plate rise to form neural folds with central neural groove. The neural folds fuse to form the neural tube. Fusion first occurs in the cervical region.
c) The **rostral and caudal neuropores** close by end of 4th week of development.
3. Complete fusion of neural folds necessary for mesodermal components to migrate dorsal to neural tube and form meninges, bone, muscle, and skin.

B. Neural Crest Cells
1. Neural crest cells are derived from neuroectoderm. They break free from lateral boundaries of neural plate during neurulation to become mesenchymal cells and migrate throughout the embryo.
2. Neural crest cells are referred to as **ectomesenchyme** because of ectodermal origin and mesenchymal cell type.
3. They contribute to the peripheral nervous system
a) Almost all peripheral ganglia, both sensory and autonomic.
b) Schwann and satellite cells ensheathing peripheral nervous system.
c) Chromaffin cells of adrenal medulla.
4. Contribute to formation of:
a) Pigment cells of skin (melanocytes).
b) Mesenchyme of head and neck.

C. Developmental Defects
1. Failure of neural folds to fuse dorsally is one of the most common and severe developmental defects of the central nervous system. Failure of the neural folds to fuse produces an extreme range of developmental defects which of necessity also includes major defects in the associated mesodermal components (meninges, neural arches, epimeric musculature, skin) which normally cover the dorsal aspect of the neural tube.
2. The spectrum of defects between **complete craniosacral myeloschisis** to **spina bifida occulta** includes **meningocele** and **meningomyelocele**.

3. **Anencephaly** is caused by failure of the neural folds to fuse in cranial region and is incompatible with extrauterine survival; it is also frequently associated with **polyhydramnios** because of absence of swallowing reflex.
4. Monitoring. Neural tube defects can usually be diagnosed in utero with **ultrasound scan**. Open neural tube defects can be detected by analyzing amniotic fluid for **alpha-fetoprotein**; levels in amniotic fluid are very high in cystic and noncystic forms.

II. PERIPHERAL NERVOUS SYSTEM: SOMATIC DIVISION

A. Somatic Motor Nerves
1. The development and distribution of efferent nervous elements in the somatic division of the peripheral nervous system is intimately associated with skeletal muscle that develops from somites or in association with branchial arches. Myogenic cells may migrate far beyond their site of origin. The course of motor nerves indicates the migratory pathway of myoblasts to reach their final destination.
2. The cell body develops from neuroectoderm of neural tube; specifically the mantle layer of the basal plate.
3. Somatic motor axons emerge from along ventrolateral sides of neural tube to form **ventral root.**
4. **Segmental emergence** of motor axons from neural tube. Result of induction of spinal nerve by segmentally organized somites. One spinal nerve for each somite.

B. Somatic Sensory Neurons
1. The primary sensory neurons for all spinal and most cranial nerves are derived from **neural crest cells**.
   a) Neural crest cells form **dorsal root ganglia** associated with mixed spinal nerves.
   b) Neural cress cells form most **sensory ganglia** associated with cranial nerves.
2. Centrally directed processes enter neural tube.
3. Peripherally directed processes.
   a) Form **dorsal root** in mixed spinal nerve.
   b) Immediately join motor fibers in mixed cranial nerves.

C. Segmental Innervation
1. Pattern of somatic innervation is **segmental**.
2. Muscles that develop from each myotome receive motor innervation via the corresponding spinal nerve.
3. Dermis that develops from each dermatome receives sensory innervation via the corresponding spinal nerve. Segmental innervation is responsible for **dermatomes**.

D. Mixed Spinal Nerves Split into 2 Branches
1. **Dorsal primary ramus** - innervates corresponding epimere.
2. **Ventral primary ramus** - innervates corresponding hypomere.

III. PERIPHERAL NERVOUS SYSTEM: AUTONOMIC DIVISION

A. Sympathetic (Thoracolumbar) Division
1. Preganglionic sympathetic fibers.
   a) Efferent fibers grow out from 14 segments of neural tube - T1-L2.
   b) Fibers connect central nervous system with sympathetic chain. These nerve fibers form the preganglionic **white rami communicantes**.

2. Ganglia.
   a) Postganglionic neurons are derived from neural crest cells.
   b) Neural crest cells migrate ventrally to reach positions adjacent to developing vertebral bodies and differentiate into neurons of **paravertebral (lateral) sympathetic chain ganglia**. Other neural crest cells continue ventral migration to anterior surface of vertebrae, surround dorsal aorta and differentiate into neurons of **prevertebral (collateral) sympathetic ganglia.**
3. Paravertebral chain ganglia.
   a) Number of chain ganglia formed on each side corresponds to 31 developing spinal nerves.
   b) Later some of the cervical, lumbar and sacral chain ganglia fuse, resulting in less lateral chain ganglia than spinal nerves.
4. Gray rami communicantes.
   a) **Gray rami communicantes provide sympathetic innervation for structures that develop in the body wall (somatopleure) and are always postganglionic.**
   b) 31 pairs of identifiable gray rami communicantes in adult. One for each spinal nerve.
5. Sympathetic innervation to thoracic and abdominopelvic viscera
   a) At cervical levels, postganglionic sympathetic nerves emerge from paravertebral ganglia and descend into the thorax as cardiac nerves to innervate the heart and lungs. In addition, in the thorax a few postganglionic sympathetic nerves emerge from cranial-most paravertebral ganglia as cardiac nerves and directly innervate the heart and lungs.
   b) At thoracic, lumbar, and sacral levels, preganglionic sympathetic fibers pass through paravertebral ganglia without synapsing and emerge from these ganglia as splanchnic nerves.
6. Prevertebral ganglia
   a) Develop from neural crest cells located around abdominal aorta.
   b) Preganglionic fibers in splanchnic nerves synapse in prevertebral ganglia.
   c) Postganglionic fibers from prevertebral ganglia innervate abdominal and pelvic viscera.

B. Parasympathetic (Craniosacral)
   1. Preganglionic parasympathetic fibers.
      a) Efferent fibers grow out from cranial or sacral (S2-4) levels of the neural tube.
      b) Fibers connect central nervous system with cranial or trunk terminal ganglia.
   2. Parasympathetic terminal ganglia are derived from **neural crest cells**.
   3. **Cranial terminal ganglia.** In the head, four terminal parasympathetic ganglia occur regularly; ciliary, submandibular, pterygopalatine, and otic. All of the cranial ganglia are constant in position, anatomically well defined, and located near organs innervated.
   4. **Trunk terminal ganglia**.
      a) Terminal ganglia are inconsistent in position, diffuse and located on or in walls of organs innervated.
      b) **Distribution of parasympathetic fibers in the trunk is limited to visceral structures originating almost exclusively from the splanchnopleure;** spinal nerves which innervate somatopleuric structures do not contain parasympathetic fibers.

c) Proximal to the descending colon, the diffuse intrinsic ganglia of the digestive tract receive their preganglionic fibers from **branches of the vagus**. Near the splenic flexure, the intrinsic ganglia receive preganglionic fibers from **pelvic splanchnic nerves**.

IV. CENTRAL NERVOUS SYSTEM

A. Neural Tube
Immediately after its formation, the neuroectodermal cells forming the wall of the neural tube are arranged in the form of a pseudostratified columnar epithelium surrounding a central neural canal. Cellular proliferation within the lateral walls of the neural tube soon results in stratification and histological and functional differentiation.

B. Development of Gray and White Matter - Histological Differentiation
1. **Ependymal layer.**
   a) Ependymal layer cells surround the neural canal.
   b) Ependymal layer cells are **mitotically active** throughout intrauterine development and are the **source of all neurons and macroglia of the central nervous system**.
   c) Ependymal layer cells that remain will form **ependyma** which surround neural canal and all of its derivatives (ventricles).
2. **Mantle layer.**
   a) Cells in mantle layer are **neuroblasts** and **glioblasts**.
   b) Neuroblasts do not divide after entering mantle layer.
   c) Neuroblasts differentiate in mantle layer into **neurons**. They will extend processes (axons) toward external limiting membrane.
   d) Mantle layer will form **gray matter** for spinal cord and brain.
3. **Marginal layer.**
   a) The marginal layer forms by **extension of axons** from mantle layer.
   b) Mantle layer thickens by differentiation of **oligodendrocytes and myelin**.
   c) Marginal layer will form **white matter** for spinal cord and brain.
4. Cortex
   a) An external layer of gray matter forms the cortex of the cerebrum and cerebellum.
   b) Neuroblasts from ependymal layer migrate superficial to marginal layer to form gray cortex.
5. Vascularization of the neural tube.
   a) Blood vessels from the developing **pia mater** penetrate and vascularize the developing neural tube.
   b) **Microglia** develop from mesoderm and appearance coincides with vascularization.

C. Development of Sensory and Motor Regions - Functional Differentiation
1. Organization of neural tube
   a) Functional differentiation is indicated by the appearance of an internal longitudinal groove in the lateral walls of the neural tube; the sulcus limitans.
   b) Sulcus limitans divides the neural tube into dorsal (alar plate) and ventral (basal plate).
2. **Alar plates** (dorsal) are concerned with sensory function and all secondary sensory neurons are located in gray matter of alar plate.

3. **Basal plates** (ventral) are concerned with motor function and all efferent neurons are located in gray matter of basal plate.

D. Differentiation of the Spinal Cord
   1. The spinal cord possesses a centrally located neural canal surrounded by ependyma, an intermediate layer of mantle cells (gray matter), and a superficial marginal layer of nerve cell processes (white matter).
   2. The gray matter (mantle) is arranged in the shape of the letter "H" to produce **anterior and posterior gray columns**. Dorsal and ventral extensions are produced by addition of mantle neuroblasts and by differentiation of nerve cell bodies.
   3. Marginal layer is organized into **dorsal, lateral, and ventral white columns**.

E. Differentiation of the Brain
   1. Three primitive brain vesicles appear in cranial neural tube.
      a) **Prosencephalon** (forebrain) - Forms secondary brain vesicles; **telencephalon and diencephalon**.
      b) **Mesencephalon** (midbrain)
      c) **Rhombencephalon** (hindbrain) - Forms secondary brain vesicles; **metencephalon and myelencephalon**.
   2. The most important gross structures developing from or associated with the definitive brain vesicles are listed below.
      a) Telencephalon
         (1) Cerebral vesicles.
         (2) Lateral ventricles.
      b) Diencephalon
         (1) Thalamus, hypothalamus, neurohypophysis, pineal gland, and retina.
         (2) 3rd ventricle.
      c) Mesencephalon
         (1) Superior and inferior colliculi and cerebral peduncles.
         (2) Cerebral aqueduct.
      d) Metencephalon
         (1) Cerebellum and pons.
         (2) 4th ventricle.
      e) Myelencephalon
         (1) Medulla.
         (2) 4th ventricle.
   3. Myelencephalon.
      a) The medulla remains structurally similar to the spinal cord. The major change is the separation of alar plates to form the dorsal lamina.
      b) Basal plate. Motor nuclei of IX, X-XI, XII.
      c) Alar plate. Secondary sensory neurons for IX, X.
   4. Metencephalon.
      a) **Cerebellum** formed by enormous development and fusion of alar plates.
      b) **Pons** formed primarily by marginal fibers crossing ependymal lamina of floor.
      c) Basal plate. Motor nuclei of V, VI, VII.
      d) Alar plate. Secondary sensory nuclei of V, VII, VIII.
   5. Mesencephalon.
      a) Retains most of basic structure of neural tube at spinal levels.

b) Basal plate. Motor nuclei of III, IV.

c) Alar plate. No secondary sensory neurons for cranial nerves. Neurons located in collicular area perform integrative functions for coordinating eye movements.

6. Diencephalon.

a) Diencephalon is considered to contain only alar components.

b) Structures derived from diencephalon include **thalamus, hypothalamus, epiphysis, neurohypophysis, and retina.**

c) Most of diencephalic vesicle becomes buried by caudal growth of telencephalon.

7. Telencephalon.

a) The telencephalon is the highest integrative center of brain and contains only the alar component of the neural tube.

b) The complex integrative functions of the telencephalon can be correlated with the presence of a multilayered neocortex (6 layers), exceedingly large bilateral hemispheres and the appearance of sulci and gyri.

F. Developmental Defects

1. Early fusion defects of the neural tube will result in spina bifida or anencephaly.

2. A wide variety of central nervous system defects may result from inadequate development of alar or basal plate derivatives at any level; e.g. cerebellar agenesis.

3. Any genetic or environmental condition which interferes with proliferative activity of germinal ependyma may result in microcephaly or mental retardation.

4. Absence of bilaterality of the forebrain may result in holoprosencephaly, arrhinencephaly, agenesis of the corpus callosum.

5. **Maternal consumption of alcohol is believed to be the major cause of congenital mental retardation.**

6. Hydrocephalus. The third and lateral ventricles become swollen with cerebral spinal fluid usually the result of constriction of the cerebral aqueduct. The cerebral cortex is abnormally thin, and the sutures of the skull are forced apart.

MUSCULOSKELETAL SYSTEM

I. AXIAL SKELETON

A. **Paraxial mesoderm**. The paraxial mesoderm forms all of the axial skeleton except the viscerocranium. It is segmented into somitomeres and somites.

1. **Somitomeres**. 7 somitomeres form in cephalic region.

2. **Somites**. 4 occipital, 8 cervical, 12 thoracic, 5 lumbar, 5 sacral, 3 coccygeal.

B. **Vertebrae and Intervertebral Disks**

1. **Sclerotome**. Under the inductive influence of notochord, the immediately adjacent portions of the somites (sclerotomes) migrate medially and fuse around the notochord to produce in the midline the segmentally arranged primary sclerotomes.

a) **Primary sclerotomes** split into cranial and caudal portions to form secondary sclerotomes.

b) **Secondary sclerotomes** form by fusion of adjacent cranial and caudal halves of primary sclerotomes. Secondary sclerotomes form definitive vertebrae and associated intervertebral disks.

c) Fusion of adjacent halves of primary sclerotomes results in one-half segmental shift. This shift results in:

          (1) **Intersegmental vessels** which cross vertebral body rather than intervertebral disk.

          (2) **Segmental muscles**, formed by the myotome, which operate across an intervertebral articulation.

          (3) **Segmental spinal nerves** which exit between vertebral bodies.

    2. **Notochord**. After inducing formation of sclerotome, the majority of the notochord degenerates. Remaining portion forms **nucleus pulposus** of intervertebral disk.

    3. Each secondary sclerotome will develop three pairs of processes:

       a) **Neural arch processes** which grow dorsomedially and fuse across midline to enclose developing spinal cord. Failure of neural arch processes to fuse results in **spina bifida**.

       b) **Transverse processes** which grow between dorsal (epimeric) and ventral (hypomeric) musculature; variable in degree of development.

       c) **Costal (rib) processes** which grow ventrolaterally into the somatopleura (primitive body wall) with other somite derivatives.

  C. **Skull**

    1. The skull can be divided into two parts: neurocranium and viscerocranium.

    2. **Neurocranium**.

       a) Develops from 7 somitomeres and 4 occipital somites.

       b) Forms protective case around the brain.

       c) Can be subdivided into two parts: cartilaginous and membranous neurocranium.

       d) **Cartilaginous neurocranium**.

          (1) **Cephalic somitomeres**. Paraxial mesodermal cells from the somitomeres migrate medially, ventral to the neural tube to participate in forming the cartilaginous neurocranium. In this location they form the parachordal, hypophyseal, trabeculae cranii, and ala orbitalis cartilages. More laterally otic capsules form around the otic vesicles. These cartilages and the occipital sclerotomes fuse and subsequently undergo endochondral ossification to form the base of the skull.

          (2) **Occipital sclerotomes**. The 4 occipital somites form sclerotomes that loose their segmentation and form the occipital bone of the skull.

       e) **Membranous neurocranium**. Paraxial mesodermal cells from the somitomeres migrate laterally and dorsally to the neural tube and undergo intramembranous ossification to form the membranous neurocranium.

    3. **Viscerocranium**. The viscerocranium develops from neural crest cells associated with the frontonasal prominence and branchial apparatus. Forms the skeleton of the face.

II. APPENDICULAR SKELETON

  A. **Lateral plate somatic mesoderm**. The skeletal elements for the limbs and limb girdles develop from lateral plate somatic mesoderm.

  B. **Endochondral ossification**. All of these bones develop by endochondral ossification, with the exception of the clavicle which develops by intramembranous ossification.

III. DEVELOPMENT OF TRUNK MUSCULATURE

A. **Myotome Origin of Trunk Musculature.** Musculature of the trunk is derived exclusively from the myotomic portions of somites. Skeletal muscle of myotomic origin is always innervated by motor neurons (basal plate) with axons leaving the spinal cord as ventral roots, i.e., **general somatic efferent**. Myotomes are responsible for the development of the segmentally arranged motor spinal nerves; the association of a myotome and its segmental spinal nerve is established very early in development and is retained through subsequent stages of migration and differentiation.

B. **Body Wall Musculature**
   1. **Dorsal or Epaxial Muscles.**
      a) Each myotome divides into a dorsal portion (epimere) which migrates dorsally and medially toward the developing neural arches and transverse processes of the vertebrae.
      b) Epaxial musculature is accompanied by its nerve supply which persists in the adult as the **dorsal primary ramus** (posterior primary division) of a spinal nerve.
   2. **Ventral or Hypaxial Muscles.**
      a) Each myotome divides into a ventral portion (hypomere) which migrates ventrally and medially within the primitive body wall (somatopleura) until it reaches the midline or **linea alba** area of the ventral body wall.
      b) Hypaxial musculature is accompanied by its nerve supply which persists in the adult as the **ventral primary ramus** (anterior primary division) of a spinal nerve.
   3. **Segmental Innervation.** In the adult, the basic segmental pattern of trunk musculature is obvious in the thoracic intercostal muscles, but segmentation is largely obscured in other areas by fusion of adjacent myotomes to form broad sheets (abdominal musculature) or long strap-like muscles (rectus abdominis). Despite these changes (fusions, splitting, etc.), the segmental innervation pattern is retained throughout life.
   4. **Definitive body wall** is formed by the migration of somite derivatives into the primitive somatopleura.

C. Appendicular Musculature
   1. **Myotomic Origin of Appendicular Musculature.** The appendicular musculature is derived exclusively from the **myotomic portions of somites**. The connective tissue (epimysium and perimysium) surrounding the myofibers and adjacent tendons develop from lateral plate somatic mesoderm.
   2. Appendicular musculature develops from ventral and dorsal condensations of somitic mesoderm.
      a) Ventral (anterior) muscle mass: gives rise to **flexors** and **pronators** of the upper limb and **flexors** and **adductors** of the lower limb.
      b) Dorsal (posterior) muscle mass: gives rise to **extensors** and **supinators** of the upper limb and **extensors** and **abductors** of the lower limb.
   3. Innervation of Appendicular Musculature.
      a) All appendicular musculature is innervated by branches of **ventral primary rami** from spinal nerves.
      b) The ventral primary rami innervating the limb and girdle musculature do not remain segmental, rather they join together at the base of the limb to form **plexi**, e.g., brachial, lumbosacral.

c) The initial segment of the combined rami form the plexus **trunks** which then divide to form anterior and posterior **divisions**. Nerves originating from the anterior and posterior divisions are subsequently distributed to their respective anterior and posterior muscle masses in the limb bud.

IV. CRANIAL MUSCULATURE

A. Cranial musculature is derived from **paraxial mesoderm**.

B. **Extrinsic Ocular Muscles.** Classically, the extrinsic ocular muscles were considered to be derived from three pairs of preotic somites. Recent evidence suggests that these skeletal muscles are derived from partially segmented cephalic paraxial mesodermal structures termed somitomeres.

C. **Branchiomeric Musculature.** Classically, the skeletal muscle that develops in association with the branchial apparatus is considered to be derived from branchiomeric mesoderm. **It should be emphasized that recent studies have suggested that branchiomeric musculature, similar to that observed for all other skeletal musculature, is derived from paraxial mesoderm; more specifically from somitomeres and the cranial-most occipital somites.**

D. **Occipital Somites.** Myotomes from the occipital somites migrate anteriorly into the region of the developing oral cavity (stomodeum and cranial foregut) and subsequently differentiate to form the tongue musculature (intrinsic and extrinsic).

V. DEVELOPMENTAL DEFECTS

A. Failure of somite(s) to develop.
   1. Failure of somites to appear during early development will produce a shortened body axis accompanied by a reduction in the number of spinal nerves. Conversely, extra somites produce an elongated body axis accompanied by an increase in spinal nerves.
   2. **Congenital scoliosis**. Individuals are born with one or more hemivertebrae (half-vertebrae).
      a) The developmental basis for forming half vertebrae can result from asymmetrical: somite formation, i.e., somites do not form in pairs; formation of primary and secondary sclerotomes; formation of the bilateral ossifications centers for the cartilaginous model of the vertebral body.
      b) Failure of a somite to appear on one side would be reflected by absence of the corresponding spinal nerve, i.e, a myotome was never present to induce nerve formation. Later regression or degeneration of a normally formed somite may produce skeletal (sclerotomic) or muscle (myotomic) defects without reducing the number of spinal nerves.

B. Failure or incomplete migration of the hypomere within the primitive body wall.
   1. The body wall remains thin and transparent because the somatopleura lacks the migratory somite derivatives; normal differentiation of the integumentary system (epidermis as well as dermis) does not occur in the absence of underlying muscle.
   2. Rupture of the persisting somatopleura is responsible for external exposure of the viscera - **thoracoschisis, gastroschisis, exstrophy of the bladder.**

INTEGUMENTARY SYSTEM

I. SKIN

    A. Skin receives major contributions from **ectoderm** and **mesoderm** germ layers.

    B. **Epidermis** is derived from the ectoderm covering the outer surface of the embryo.

    C. **Dermis** or corium is derived from mesoderm.
        1. Dermis is derived from the dermatomic portion of the somite, somatic layer of the primitive body wall (somatopleura), and in the head neural crest-derived ectomesenchyme.
        2. Normal differentiation of the epidermis and general integumentary structures (hair and glands) appears to occur only in the presence of underlying skeletal muscle.

II. GENERAL INTEGUMENTARY COVERING

    A. Epidermis
        1. Initially, the ectodermal epithelium covering the embryo is composed of a simple cuboidal epithelium. A second outer layer of cells termed **periderm** is formed and later lost. The deeper layer of cuboidal cells will form the **basal cell layer** (stratum basale) of the epidermis.
        2. An intermediate layer of cells accumulates below the periderm which forms three layers of **keratinocytes**: stratum spinosum, stratum granulosum and stratum corneum.
        3. Vernix caseosa: is a white fatty secretion product produced by the fetal sebaceous glands; it is particularly thick over the scalp, back and skin creases around joints.

    B. Pigment cells
        1. Neural crest cells appear in the deepest layer of the developing epidermis (basal cell layer) during the first month of fetal life and subsequently complete their differentiation into **melanocytes**.
        2. Pigmentation (moles) and vascular (angiomas) malformations are minor development anomalies and can be found in almost any body area; most are small and insignificant. Occasionally, however, these malformations may be quite large and are disfiguring when present on the face, scalp or neck, e.g., deeply pigmented and elevated nevi, port wine stain.

III. SPECIALIZED INTEGUMENTARY STRUCTURES

    A. General Development. All of the specialized integumentary structures (glands, hair, teeth, nails) are formed by the same basic developmental process, i.e., proliferation of the basal cell layer of the epidermis into the underlying mesenchyme.

    B. Mammary Glands. Mammary glands are the most characteristic feature of mammals and are the first specialized integumentary structures to appear.
        1. **Mammary ridge**: an ectodermal band that appears on the anterior body wall during the sixth week of embryonic life and extends caudally from the axillary to the inguinal region. In humans, mammae are restricted to the pectoral region and as a consequence, portions of the mammary line cranial and caudal to this area regress.

2. **Lactiferous ducts**: in the pectoral region, 16-24 solid cords of epithelial cells appear within a small circumscribed (nipple) area and continue to grow and branch within the adjacent mesenchyme. Canalization occurs during the latter half of pregnancy. The mammary gland of the newborn consists primarily of proximal ducts with a few rudimentary acini.

3. **Witch's milk**: a secretory product of the newborn mammae, is thought to be produced as a result of high levels of fetal steroids (estrogen and progesterone) and placental lactogen. The rudimentary acini regress and secretion ceases as hormone levels decline following parturition.

4. Developmental defects.
    a) **Polythelia** or extra nipples is very common and may occur as elevated or pedunculated pigmented 'moles' anywhere along the mammary line, i.e., axilla to medial surface of the thigh.
    b) **Polymastia** or extra mammary glands is very rare but they have been reported from the axilla to the medial surface of the thigh.
    c) **Amastia** or complete absence of a mammary gland on one or both sides is so rare that the defect is almost nonexistent.

C. Teeth
1. **Dental laminae** for the upper and lower jaws appear in the stomodeal (ectodermal) portion of the oral cavity during the seventh week of development; the two laminae will subsequently form enamel organs for all of the deciduous (20) and permanent (32) teeth.

2. **Ameloblasts** of the ectodermally derived enamel organ will secrete only the enamel covering for the crown of the tooth.

3. **Odontoblasts** originating from the neural crest ectomesenchyme of the dental papilla will form the dentin for all remaining parts of the tooth, i.e., dentin supporting the enamel and forming the roots of the tooth.

4. The enamel organs for the 32 permanent teeth begin to appear during fetal life, however their formation is not complete at birth. The enamel organs for the last permanent molars (wisdom teeth) develop at about the age of five years. Tetracycline, which localizes in mineralizing tissues, should be used with caution in pregnant women, infants and young children.

5. **Developmental defects**. Some abnormalities in tooth morphology and number are familial traits; most if not all, appear to be inherited as autosomal dominants.

D. Hair
1. **Hair follicles** first appear on the head and face during the third month.
2. **Lanugo hair** is very fine and dense hair which covers the developing fetus. It is normally shed before or shortly after birth.

E. Sebaceous glands
1. Primordia for sebaceous glands usually develop from the epithelial cells of hair follicles and are, therefore, somewhat later to appear (4-5 months) than the other integumentary structures.
2. The central cells of the gland primordia undergo fatty degeneration of the **holocrine** type to form the fatty component of the vernix caseosa.

F. Sweat glands: unbranched epithelial cords forming the sweat gland primordia appear during the third month and canalize a short time later.

G. Nails
    1. The plate-like epithelial folds forming the primordia for nails appear near the tips of the digits during the third month. Nail folds migrate to a more proximal location on the dorsal aspect of the digits; dorsal migration may explain why the nails and adjacent skin are innervated by nerves derived from ventral cutaneous nerves.
    2. Nails grow slowly during fetal life and reach the tip of the digits about the time of parturition.

DEVELOPMENT OF THE HEAD AND NECK

I. ROLE OF BRANCHIAL APPARATUS IN HEAD AND NECK DEVELOPMENT

    A. **Branchial apparatus** is a series of structural complexes (branchial arches) which surround the pharynx.

    B. Evolutionary Importance. The branchial apparatus in primitive chordates gives rise to the gills which are associated with feeding and respiration.

    C. Human Development. Branchial apparatus is present during early development. The branchial apparatus undergoes a dramatic process of morphogenesis. Almost all of the structures found in the adult human head and neck are associated with the development and differentiation of the branchial arches.

II. FORMATION OF BRANCHIAL APPARATUS AND STOMODEUM

    A. Head Folding
        1. The oropharyngeal membrane and cardiogenic region, which are cranial to the notochord, are "swung" ventrally and caudally.
        2. The foregut is formed as a result of head folding.

    B. Branchiomeric Mesenchyme
        1. **Cephalic neural crest ectomesenchyme**. Most of the branchiomeric mesenchyme originates from cephalic neural crest cells. These cells migrate laterally and ventrally to the pharynx where they rapidly proliferate and replace the lateral plate mesoderm. These cells form the skeleton and connective tissue that develops from the branchial apparatus.
        2. **Cephalic paraxial mesoderm**. Cephalic paraxial mesoderm is organized into somitomeres. Mesodermal cells, equivalent to the myotome that forms from somites, migrates along with neural crest cells to form the branchial apparatus. These cells will give rise to **branchiomeric musculature**.

    C. Stomodeum
        1. **Stomodeum** is ectodermal lined depression that forms as result of proliferation of mesenchyme around the circumference of the oropharyngeal membrane.
        2. **Oropharyngeal membrane** initially separates stomodeum and foregut (pharynx). Later will degenerate allowing communication between stomodeum and foregut.
        3. Stomodeum will form the anterior 2/3 of the definitive oral cavity. The posterior 1/3 develops from foregut.

4. **Rathke's pouch**. An ectodermal outgrowth from the roof of the stomodeum that forms the primordium of the **adenohypophysis**.

III. ORGANIZATION OF THE BRANCHIAL APPARATUS

A. Branchial Arches
   1. Branchial apparatus is composed of a series of swellings, **branchial arches**, that occur around the pharynx.
   2. Six branchial arches develop.
      a) Branchial arches are numbered in craniocaudal sequence.
      b) The fifth arch is rudimentary and transitory and not considered to give rise to any adult structures.
      c) The first arch is separated into two paired prominences, **maxillary prominences** and **mandibular prominences**.
   3. Branchiomeric mesenchyme. Each branchial arch consists of a central core of branchiomeric mesenchyme containing 4 structural components.
      a) **Paired aortic arches**. These participate in formation of the definitive arterial vasculature in the root of the neck; aorta, common carotids, right subclavian, and pulmonary arteries.
      b) **A cartilaginous bar**. All branchial arch cartilages give rise to adult skeletal structures. Branchial arch cartilages contribute to the viscerocranium, neurocranium (a small portion), hyoid bone and laryngeal cartilages.
      c) **A cranial nerve**. Each branchial arch receives a specific cranial nerve that will provide afferent and efferent innervation for all derivatives of that specific branchial arch.
      d) **A muscular component**. Each branchial arch receives myoblasts from cephalic paraxial mesoderm (somitomeres). Branchiomeric muscles receive their innervation from cranial nerves that go to a specific branchial arch. The musculature developing in association with the branchial apparatus receives special visceral efferent innervation.
   4. Boundaries of branchial arches. Each arch is delineated by **branchial grooves** and **pharyngeal pouches** which, like the arches, are numbered; both grooves and pouches are located caudal to the arch with the corresponding number. The caudal limits of the sixth arch are not indicated by the presence of grooves and pouches and as a consequent it is directly continuous with the post-branchial body wall.

B. **Branchial Grooves**. There are four external, ectoderm-lined, grooves separating each branchial arch.

C. **Pharyngeal Pouches**. There are four internal, endoderm-lined, evaginations from the pharynx separating each branchial arch.

D. **Branchial Membrane**. There are four regions between each branchial arch where the ectoderm of the branchial groove and the endoderm of the pharyngeal pouch become closely apposed.

IV. FATE OF THE BRANCHIAL GROOVES

A. Ectodermal Epithelium. The ectodermal epithelium of the stomodeum, branchial arches and grooves is the presumptive integumentary epithelium for the future oral cavity and skin. All of the ectodermal epithelial areas will receive their general somatic afferent innervation from the nerves of the arches involved, i.e., the stomodeal portions of the nasal cavities and anterior oral cavity receive these fibers from the maxillary (V₂) and mandibular (V³) divisions of the trigeminal nerve.

B. Branchial Groove 1
   1. **External auditory canal**. The first branchial groove persists as the external auditory canal.
   2. **Tympanic membrane**. The first branchial membrane persists to form the tympanic membrane.
   3. **Innervation**. The general somatic afferent innervation for this area is derived from the nerves of the adjacent arches, i.e., trigeminal and facial, with minor contributions from glossopharyngeal and vagus nerves.

C. Branchial Grooves 2, 3, and 4
   1. Regression of branchial grooves 2-4. All branchial grooves caudal to the first are obliterated.
   2. **Cervical sinus**. During regression of the branchial grooves a deep ectodermal-lined depression, the cervical sinus, forms on the lateral side of the neck connecting with deep lying branchial grooves. Normally the cervical sinus is lost and the neck obtains its smooth adult contour.

V. FATE OF THE BRANCHIOMERIC MESENCHYME

A. Branchial Arch 1
   1. Skeleton - **Meckel's cartilage**
      a) Forms the **malleus and incus** by endochondral ossification, fibrous anterior malleolar and sphenomandibular ligaments, and the fibrous core of mandible.
      b) First arch mesenchyme will also form skeletal elements by intramembranous ossification.
         (1) Maxillary prominence forms **maxilla, zygomatic and temporal squama**.
         (2) Mandibular prominence forms **mandible**.
   2. Cranial nerve.
      a) **Maxillary division of trigeminal nerve (V²)** innervates maxillary prominence.
      b) **Mandibular division of trigeminal nerve (V³)** innervates mandibular prominence.
   3. Muscles.
      a) Muscles only develop from mesenchyme in mandibular prominence.
      b) Forms muscles of mastication (temporalis, masseter, and medial and lateral pterygoids), mylohyoid, anterior belly of digastric, tensor tympani, and tensor veli palatini.

B. Branchial Arch 2
   1. Skeleton - **Reichert's cartilage**. Forms stapes, styloid process, lesser cornu, and upper half of hyoid bone by endochondral ossification and the stylohyoid ligament.
   2. Cranial nerve. **Facial nerve (VII)** innervates second branchial arch.
   3. Muscles.
      a) Forms muscles of facial expression which migrate in superficial fascia throughout head, neck and antero-superior thorax.
      b) Forms stapedius, stylohyoid, and posterior belly of digastric.

C. Branchial Arch 3
1. Skeleton - Third arch cartilage. Forms greater cornu and lower half of hyoid bone by endochondral ossification.
2. Cranial nerve. **Glossopharyngeal nerve (IX)** innervates third branchial arch.
3. Muscles. Forms stylopharyngeus.

D. Branchial Arch 4
1. Skeleton - Fourth arch cartilage. Forms laryngeal cartilages.
2. Cranial nerve. **Superior laryngeal branch of vagus nerve (X)** innervates fourth branchial arch.
3. Muscles. Forms constrictors of pharynx, cricothyroid, levator veli palatini

E. Branchial Arch 6
1. Skeleton - Sixth arch cartilage. Forms laryngeal cartilages.
2. Cranial nerve. **Recurrent laryngeal branch of vagus nerve (X)** innervates sixth branchial arch.
3. Muscles. Forms intrinsic muscles of larynx.

VI. FATE OF THE PHARYNGEAL POUCHES

A. Pharyngeal Pouch 1
1. The first pouch forms the **pharyngotympanic tube** (Eustachian), **cavity** of the middle ear and inner mucosal layer of the **tympanic membrane**.
2. Dorsal expansion of the first pouch surrounds the bones developing from the dorsal ends of Meckel's (malleus and incus) and Reichert's (stapes) cartilages and establishes the definitive relationship of ear ossicles traversing the cavity of the middle ear.

B. Pharyngeal Pouch 2. The second pouch forms the epithelium and crypts of the **palatine tonsil**; the lymphocytes will migrate into this area later.

C. Pharyngeal Pouch 3
1. The third pouch forms the **inferior parathyroids** and the endodermal primordium for the **epithelial reticulum** of the definitive thymus; lymphocytes begin to invade the thymic epithelial cords at about nine weeks.
2. These structures lose their association with the pharynx and migrate caudally.

D. Pharyngeal Pouch 4
1. The fourth pouch forms the **superior parathyroids** and **ultimobranchial body**.
2. The superior parathyroids break free from the pharynx and migrate caudally.
3. The parafollicular cells develop in association with the ultimobranchial body and invade the developing thyroid gland.

VII. BRANCHIAL MALFORMATIONS

A. Branchial sinus, cyst or fistula
1. These malformations are associated with maldevelopment of branchial grooves 2-4 and the cervical sinus.
2. They are located in the side of the neck just under or anterior to the sternocleidomastoid.

B. First arch syndrome
1. This syndrome is the result of insufficient migration of neural crest cells into the first branchial arch.
2. Malformations include those of the ear, mandible, zygoma, and palate.

C. DiGeorge Syndrome
1. The third pharyngeal pouches do not develop properly.
2. This syndrome is characterized by thymic agenesis with impaired cell mediated immunity and with hypoparathyroidism and tetany.
3. Associated heart malformations include abnormalities in division of truncus arteriosus and bulbus cordis.

D. Ectopic Glandular Tissue
1. Any of the glands (thymus, superior parathyroids and inferior parathyroids) may be ectopic in location.
2. This is the result of abnormal migration away from the pharynx.

DEVELOPMENT OF THE FACE, PALATE, TONGUE AND THYROID GLAND

I. DEVELOPMENT OF THE FACE

A. Five Facial Primordia. All of the facial primordia surround the stomodeum and form the boundaries of the primitive oral opening; these primordia are the: **frontonasal prominence**, **maxillary prominences** and the **mandibular prominences**.
1. Frontonasal Prominence. The superior boundary of the stomodeum is formed by the unpaired frontonasal prominence; it will ultimately form facial areas above the external nares and tip of the nose. Most of the structures developing from the frontonasal prominence will be innervated by the **ophthalmic division** (V^1) of the trigeminal nerve.
2. Maxillary Prominences. The lateral boundaries of the stomodeum are formed by the paired maxillary prominences of the first branchial arch; they will ultimately form most of the facial areas between the external nares and superior boundary of the definitive oral opening, i.e., upper jaw. The **maxillary division** (V^2) of the trigeminal nerve will supply almost all of the structures developing from the maxillary portion of the first branchial arch.
3. Mandibular Prominences. The inferior boundary of the stomodeum is formed by the paired mandibular prominences of the first branchial arch; they will ultimately form facial areas below the definitive oral opening. Structures originating from this primordium will be innervated by the **mandibular division** (V^3) of the trigeminal nerve.

B. Differentiation of the Frontonasal Prominences
1. Superior (frontal) portion. Forms the **frontal** (forehead) and **interorbital areas** of the face and the **dorsum of the nose**; near the midline, significant contributions are also made to the supra- and infraorbital areas.
2. Inferior (nasal) portion. Forms the **nasal placodes** with their associated **medial and lateral nasal prominences**; subsequent differentiation of the lateral and medial nasal prominences and formation of the upper jaw requires contact and fusion with the subjacent **maxillary prominences**.
   a) Nasal placodes.
      (1) **Nasal placodes** originate as localized thickenings in the superficial ectoderm immediately above the nasolacrimal groove separating the frontonasal and maxillary prominences. A short time later, accumulations of mesenchyme

around the periphery of the placodes elevates the adjacent ectoderm above the placode to produce the **nasal pit**.

(2) Primary sensory neurons differentiate in the placode epithelium to form **olfactory neurons**. Centrally directed processes grow into the underlying telencephalon and make synaptic connections with secondary sensory neurons developing in the **olfactory lobes**.

b) Lateral nasal prominences.

(1) Proliferation of mesenchyme enclosing the superior and lateral aspects of the nasal pit.

(2) Maxillary prominences grow toward the midline bringing the maxillary and lateral nasal prominences into apposition. After contact and fusion, the lateral nasal prominences form the sides of the nose, i.e., **nasal alae**.

c) Medial nasal prominences.

(1) Proliferation of mesenchyme enclosing the superior and medial aspects of the nasal pit.

(2) Medial nasal prominences merge together to form the right and left halves of the very prominent **median nasal prominence**. Continuing growth of the maxillary prominences toward the midline brings them into contact with the centrally located median nasal prominence. After fusion the median nasal and maxillary prominences form the central and lateral areas of the upper jaw primordium.

C. Differentiation of the Median Nasal and Maxillary Prominences
1. Superficial contributions. The labial or superficial portion of the median nasal prominence forms the **philtrum** or **central area** of the upper lip while **lateral areas** are formed by the superficial part of the maxillary prominences.
2. Deep contributions.
   a) Median nasal prominence.
      (1) **Medial portion of the maxilla** (premaxilla) which is associated with incisor teeth, i.e., the intermaxillary portion of the adult human maxilla.
      (2) **primary palate** (median palatine process) and the nasal septum.
   b) Maxillary prominence.
      (1) **Lateral portion of the maxilla** (maxilla proper) which is associated with the canine and post-canine (premolars and molars) teeth.
      (2) **Secondary palate** (lateral palatine processes).

D. Differentiation of the Mandibular Prominences. The mandibular prominences of the first branchial arch fuse across the midline to form **primordium of the lower jaw** and inferior boundary of the definitive oral opening.

E. Facial Musculature
1. Muscles of facial expression. The frontonasal prominence and first branchial arch (maxillary and mandibular prominences) are secondarily invaded by **myoblasts originating from the second arch**. These migratory myoblasts retain their innervation by the nerve to the second branchial arch, the **facial nerve (VII)**.
2. Muscles of mastication. The branchiomeric mesenchyme of the mandibular prominence differentiates into the **masticatory muscles** and retains it innervation from the **mandibular division (V^3) of the trigeminal nerve**.

F. Facial Development During Fetal Life. Complete facial development occurs slowly and is effected primarily by changes in the proportion and relative positions of the five primordia.

Enlargement of the developing brain (especially the telencephalon) produces a prominent forehead and appears to be instrumental in moving the laterally placed eyes to their definitive frontal position. Differentiation of the mandibular prominences forming the lower jaw appears to elevate the ears.

II. DEVELOPMENT OF NASAL CAVITIES

A. Nasal pits and sacs. Ectoderm-lined depressions that result from proliferation of mesenchyme in lateral and medial nasal prominences.

B. Oronasal membrane.
   1. Initially separates paired nasal cavities from oral cavity.
   2. Rupture of oronasal membrane. Rupture of membrane allows continuity between paired nasal cavities and oral cavity.

III. DEVELOPMENT OF THE PALATE

A. The **definitive palate** separating the dorsal (respiratory) and ventral (digestive) passages at stomodeal levels is formed by fusion of the **primary and secondary palates**.
   1. Primary palate is a wedge-shaped structure originating from the deep portion of the median nasal prominence, i.e., the **palatine process**.
   2. Secondary palate is formed by fusion of the **lateral palatine processes** originating from the deep portion of the maxillary prominences.

B. Fusion of palatine process and lateral palatine processes. The initial fusion occurs rostrally between the right and left sides of the **median palatine process** and the cranial ends of the **lateral palatine processes**; the fusion then progresses caudally to meet at the apex of the wedge, i.e., future incisive area. Subsequent fusions caudal to the incisive area involves only the lateral palatine prominences and the inferior edge of the nasal septum; the **palatine raphe** permanently marks the fusion site of the lateral palatine process. In the adult, the **incisive papilla** indicates the approximate boundary between the primary and secondary palates; in macerated skulls, the landmark is the **incisive foramen**.

C. Developmental Defects.
   1. Anterior cleft malformations - Cleft lip
      a) Results from incomplete fusion of **maxillary prominence** with **medial nasal prominence**.
      b) These malformations may be classified.
         (1) Incomplete or complete cleft lip depending on whether only superficial or superficial and deep structures are involved.
         (2) Unilateral or bilateral cleft lip.
   2. Posterior cleft malformations - Cleft palate
      a) Results from incomplete fusion of **lateral palatine processes** with each other, **median nasal septum** and **median palatine process**.
      b) These malformations may be classified.
         (1) Incomplete or complete cleft palate depending on extent lateral palatine processes fail to fuse.
         (2) Unilateral of bilateral cleft palate.
   3. Anterior and posterior cleft malformations may occur separately or may occur together.

IV.   DEVELOPMENT OF THE TONGUE

   A.   Tongue Development
      1.   Anterior two-thirds.
         a)   Median tongue bud. Median swelling in **first branchial arch** lining stomodeum.
         b)   Distal tongue buds. Develop lateral to median tongue bud on first branchial arch. Overgrow median tongue and give rise to **anterior two-thirds of tongue.**
      2.   Posterior one-third.
         a)   Copula. Median swelling of **second branchial arch**.
         b)   Hypobranchial eminence. Large median swelling of **third and fourth branchial arches**. Overgrows copula to form **posterior one-third of tongue.**

   B.   Innervation
      1.   The mucosa for the anterior two-thirds of the tongue develops from the stomodeal ectoderm of the first branchial arch. It receives its general somatic afferent innervation from the **mandibular division of the trigeminal nerve (V^3).**
      2.   Because of the overlapping pattern for general and special afferent innervation in the oropharyngeal mucosa, the first arch area receives its special taste fibers from the **facial nerve (VII) via the chorda tympani.**
      3.   The mucosa of the posterior one-third of the tongue develops from the pharyngeal endoderm of the third arch (cranial part of hypobranchial eminence); it receives general visceral afferent fibers and special visceral afferent fibers for taste from the **glossopharyngeal nerve (IX).** The **vagus nerve (X)** innervates the most posterior part of the tongue.

   C.   Musculature
      1.   Occipital myotomes. Myoblasts migrate from **occipital myotomes** into the developing tongue to form all the intrinsic and extrinsic muscles.
      2.   Innervation. The **hypoglossal nerve (XII)** follows the migrating occipital myoblasts and supplies their general somatic efferent innervation.

V.   DEVELOPMENT OF THE THYROID GLAND

   A.   Development
      1.   Thyroid primordium develops as a midline endodermal invagination in the floor of the pharynx at the **foramen cecum.**
      2.   Descends at end of **thyroglossal duct** in midline of neck to reach adult location.
      3.   Thyroglossal duct degenerates.

   B.   Developmental Defects
      1.   Thyroglossal duct may persist as midline thyroglossal cyst or sinus.
      2.   Ectopic thyroid gland. Thyroid gland may not descend to normal adult location.

RESPIRATORY SYSTEM

I.   DEVELOPMENT OF TRACHEA AND LARYNX

   A.   Separation of Respiratory Primordium From Pharyngeal Foregut
      1.   Laryngotracheal groove. The primordium for the lower portion of the respiratory system appears during differentiation of the branchial apparatus and is first seen as a midline **laryngotracheal groove** in the floor of the pharyngeal foregut.

2. Laryngotracheal diverticulum. The caudal portion of the laryngotracheal groove expands ventrally to form the **laryngotracheal diverticulum**. Continued ventral expansion forms the lung bud.

3. Tracheoesophageal folds. Separation of the respiratory primordium from the pharyngeal foregut is the result of **laryngotracheal folds** which subsequently fuse across the midline to form a **tracheoesophageal septum**.

4. Laryngotracheal tube. The tracheoesophageal septum divides the cranial part of the foregut into a ventral portion, the **laryngotracheal tube** (primordium of the larynx, trachea, bronchi, and lungs), and a dorsal portion (primordium of the oropharynx and esophagus).

B. Development of the Larynx
1. Glottis. The communication between the laryngotracheal groove and pharyngeal lumen persists as the laryngeal orifice or **glottis**.
2. **Epiglottis** is formed by the adjacent **hypobranchial eminence** (caudal part)
3. Laryngeal cartilages are formed by fourth and sixth arch cartilages, except for the epiglottis.

C. Developmental Defects.
1. **Tracheoesophageal fistula** involves abnormal communication between esophagus and trachea. Most common type is associated with esophageal stenosis or atresia. Results from defective formation of the tracheoesophageal septum.
2. Associated conditions may be present.
   a) Problems associated with feeding.
   b) **Polyhydramnios**. Excess amniotic fluid due to problems with swallowing amniotic fluid.

II. DEVELOPMENT OF BRONCHI AND LUNGS

A. Lung Bud.
1. Endodermal lined diverticulum surrounded by splanchnic mesoderm grows ventrally and caudally from **laryngotracheal diverticulum**.
2. Divides into two bronchial buds.

B. Bronchial Buds.
1. Form **primary bronchi**.
2. Lateral growth of the bronchial buds. They project into the pleural portion of the embryonic coelom and carry with them a layer of splanchnic mesoderm which forms the **visceral pleura** and other stromal elements for the lung.

C. Subdivision of Primary Bronchi.
1. Primary bronchi subdivide into secondary bronchi.
2. Secondary bronchi undergo repeated divisions to produce tertiary or segmental bronchi.
3. Tertiary bronchi continue to divide to form respiratory bronchioles.

III. DIFFERENTIATION OF THE LUNGS

A. Stages of Lung Development. Lung development is usually divided into four stages which may overlap by several weeks because differentiation in the apical portions of the lung occurs earlier than comparable changes in the basal areas.

1. **Glandular or Pseudoglandular Period** (weeks 5-17). Only the conducting system develops during this stage, i.e., through the terminal bronchioles.
2. **Canalicular Period** (weeks 13-25). Luminal diameter of the conducting system increases and development of respiratory areas (respiratory bronchioles) begins. At the end of this period, a few terminal sacs with flattened epithelium (**primitive alveoli**) begin to appear.
3. **Terminal Sac Period** (weeks 24 to birth). As the name implies, this period is characterized by the development of large numbers of **terminal sacs** and is accompanied by a marked increase in vascularity. By the 28th week of development, the respiratory area and vascularity are adequate for survival of premature infants. Surface area for gaseous exchange and vascularity appear to be more important to survival than epithelial flattening, i.e., differentiation of **alveolar type I** epithelial cells. Surfactant producing **alveolar type II** cells begin to appear at about 28 weeks. Terminal sacs or 'primitive alveoli' are considered to correspond to the alveolar ducts of the adult lung.
4. **Alveolar Period** (late fetal to 8 years). During infancy and early childhood, the number of primitive alveoli is increased by distal proliferation as the more proximal areas differentiate into **alveolar ducts and mature alveoli**. The rate of proliferation declines during late childhood when the rate of maturation begins to exceed the rate of proliferation. Unlike primitive alveoli, the mature alveoli cannot proliferate to form new generations of alveoli.

B. Developmental Defects. Although prematurity is not a developmental defect, the high mortality rate in premature infants is frequently associated with respiratory problems. The single most crucial period occurs during terminal sac formation (Stage 3). In infants delivered before the 28th week, the vascularity and surface area available for gaseous exchange are usually inadequate for survival. The chances for survival increase after the 28th week but may be complicated by **hyaline membrane disease or respiratory distress syndrome of the newborn**. An important contributing factor in development of hyaline membrane disease is a deficiency or absence of **pulmonary surfactant**. In pregnancies with complications, lung maturation can be monitored by surfactant levels in the amniotic fluid.

DIGESTIVE SYSTEM AND MESENTERIES SYSTEM

I. PRIMITIVE GUT

A. Basic Organization of Gut Tube
  1. The digestive system acquires its basic tubular configuration as a result of body folding. The primitive **foregut**, **midgut** and **hindgut** regions are identifiable by the fourth week of development.
  2. The endoderm of the primitive gut will persist to form the epithelial **parenchyma** for all segments of the definitive gut, for all parts of the respiratory system caudal to the stomodeum, and for the accessory digestive glands (liver and pancreas). **Stromal elements** for all of these structures will differentiate from the associated layer of splanchnic mesoderm.
  3. The most cephalic levels of the oral cavity and caudal levels of the anal canal do not originate from the primitive gut. The ectodermal epithelium in these areas is derived from the **stomodeum** and **proctodeum**, respectively.

4. Dorsal Mesentery. After coelom formation and development of the body folds, the tubular gut is suspended from the dorsal midline of the body wall by double layers of splanchnic mesoderm which form the **primitive dorsal mesentery**.

5. Ventral Mesentery. A ventral connection between the tubular gut and anterior body wall is found only at the caudal or venous end of the embryonic heart; this 'primitive ventral mesentery' is the important developmental landmark known as the **septum transversum**.

B. Differentiation of the Foregut
1. Foregut derivatives. In craniocaudal sequence, the digestive portion of the foregut forms the posterior one-third of the oral cavity, oropharynx, esophagus, stomach and upper half of the duodenum including the common bile duct and liver as well as the pancreatic ducts and pancreas (endocrine and exocrine).

2. Vasculature. Definitive foregut areas are supplied by branches from the external carotids, small vessels arising directly from the thoracic aorta and the celiac artery.

3. Developmental defects. Due to the common origin from the foregut of the trachea and esophagus, defective partitioning by the laryngotracheal folds and/or the tracheoesophageal septum may give rise to stenotic and/or atretic segments in one or both structures or to fistulous connections between the trachea and esophagus.

C. Differentiation of the Midgut
1. Gut rotation.
   a) The midgut loop, when viewed from the ventral aspect, undergoes a single, **counterclockwise rotation of approximately 270 degrees**. The axis for rotation is formed by the superior mesenteric artery and stalk of the yolk sac. Complete rotation occurs over a period of several weeks.

   b) Herniation phase. The midgut loop herniates into the **umbilical (extraembryonic) coelom** around the sixth week of development, presumably due to rapid growth of gut tube. Within the umbilical coelom, the midgut loop undergoes an **initial rotation of 90 degrees**, i.e., cranial limb to the right, caudal limb to the left.

   c) Return phase. During the tenth week the intestines return to the abdomen. The cranial limb begins to retract first and undergoes a **counterclockwise rotation of 180 degrees** to pass down and under the rotational axis (superior mesenteric artery) and under the more slowly developing caudal limb. The return phase is completed when the lagging caudal limb has returned to the abdominal cavity but gut rotation is not complete until all of the visceral structures have reached their definitive positions. It should be noted that after completion of the return phase, the cecum is still located near the liver in the upper right quadrant. Continuing growth of adjacent areas during the post-return period allows the cecum to descend and eventually reach its definitive location. Gut rotation is complete only when the cecum is positioned in the lower right quadrant of the abdominal cavity.

   d) After gut rotation, the original midline attachment of the primitive dorsal mesentery will be altered by secondary fusions to produce the attachment sites seen in the definitive dorsal mesentery of the adult; the latter process is referred to as **gut fixation**. During fixation, the mesentery of the gut segment being "fixed" is lost by fusion with the peritoneum covering adjacent areas of the body wall and in the process, the corresponding gut segment usually becomes **retroperitoneal**.

2. Midgut derivatives. The definitive midgut forms the duodenum below the origin of the common bile duct, the jejunum, ileum, cecum, appendix, ascending colon and transverse colon to a point near the splenic flexure; all derivatives begin their differentiation during gut rotation.

3. Vasculature. Definitive midgut areas are supplied by the **superior mesentery artery**; some overlap with foregut vasculature (celiac) and hindgut (inferior mesenteric) occurs in junctional areas.

4. Developmental defects.

    a) **Omphalocele**. The presence of abdominal viscera in a persistent umbilical coelom covered by amnion produces an omphalocele and may be the result of incomplete return of the midgut loop.

    b) Malrotation. Results from incomplete or clockwise rotation. Complications include duodenal obstruction, volvulus and intussusception.

    c) Yolk stalk defects.

        (1) Remnants of the yolk stalk (**Meckel's diverticulum**) are found in about 2 percent of the adult population and are usually asymptomatic; they are always located on the antimesenteric border of the ileum about 18-25 inches above the ileocecal junction.

        (2) Persistence of the entire yolk stalk produces a connection between the terminal ileum and anterior body wall at the umbilicus; the connection may be fibrous, fibrocystic or fistulous if completely canalized.

    d) Stenosis, atresia or lumenal duplications. Because the gut lumen is normally occluded during early stages of development, stenosis, atresia and luminal duplications can occur if recanalization fails-to occur or if the recanalization process is abnormal. These occur most frequently in areas with small luminal diameters during development, esophagus or duodenum. In newborn frequently associated with projectile vomiting; a clue to the exact level of obstruction may be provided by the presence or absence of bile stained vomitus. Such obstructions may be associated with polyhydramnios.

    e) Pyloric sphincter hypertrophy. Thought to have a genetic basis and is more prevalent in males.

D. Differentiation of the Hindgut

  1. Cloaca.

    a) The terminal position of the primitive hindgut is expanded to form the **cloaca**. The cloaca forms the terminal portion of the digestive, urinary and reproductive passages.

    b) Urorectal septum. Development of the **urorectal septum** and its subsequent fusion with the **cloacal membrane** divides the common cloacal chamber into dorsal (**rectal**) and ventral (**urogenital**) cavities; the same fusion also divides the cloacal membrane into dorsal (anal) and ventral (urogenital) membranes.

    c) Proctodeum. Mesodermal cells will accumulate around the periphery of the anal membrane to produce the ectodermal-lined anal pit or **proctodeum**; subsequent degeneration of the membrane produces the definitive anal opening and the lower one-third of the anal canal.

  2. Hindgut derivatives. The digestive portion of the hindgut contributes to the formation of the descending colon, sigmoid colon, rectum and upper two-thirds of the anal canal;

the lower one-third develops from the proctodeum. The junction of endodermal and ectodermal areas is indicated by the **pectinate-line** at the level of the **anal valves**.

3.   Vasculature. Digestive derivatives of the hindgut are supplied by the **inferior mesenteric artery** with additional contributions to the lower rectum from the middle (internal iliac) and inferior (internal pudendal) rectal arteries; some overlap with vasculature of the midgut (superior mesenteric) occurs near the splenic flexure.

4.   Developmental defects.

   a)   Anorectal anomalies. Most **anorectal anomalies** result from abnormal development of the urorectal septum. Abnormalities in formation and fusion of the urorectal septum may result in **fistulous connections** between the rectum and urogenital sinus with or without abnormal terminations of the urinary and genital ducts. Failure of the anal membrane to rupture results in **imperforate anus**.

   b)   Congenital megacolon (**Hirschsprung's disease**) is caused by defective development of the myenteric plexus and is attributed to failure of **migratory neural crest cells** to invade the developing musculature. Absence of peristaltic movement in the involved segments creates an obstruction resulting in the accumulation of gut contents in normal gut above the lesion.

E.   Accessory Digestive Structures

1.   Pancreas.

   a)   The pancreas develops from **dorsal and ventral evaginations** from the foregut endoderm. The larger and more cephalic **dorsal bud** grows cranially and extends into the developing greater omentum. The smaller and more caudally located **ventral bud** turns dorsally to reach the duodenal mesentery and fuses with the dorsal outgrowth. After fusion, the duct of the ventral outgrowth becomes the terminal portion of the **main pancreatic duct** while that of the dorsal outgrowth persists as the **accessory pancreatic duct**.

   b)   Derivatives. The dorsal bud forms almost all of the adult pancreas (upper half of the head, neck, body, tail). The ventral bud forms the lower part of the head and uncinate process. The endocrine portion of the pancreas (islets of Langerhans) proliferates as solid epithelial cords from the duct system of the exocrine pancreas. Insulin secretion begins about the middle of gestation.

   c)   Vasculature. The pancreas receives its arterial blood supply via branches from the **celiac and superior mesenteric arteries**. Despite its dual blood supply the pancreas is considered a foregut derivative.

2.   Liver, gall bladder and biliary ducts.

   a)   The liver arises as a ventral evagination (**hepatic diverticulum**) from the most caudal level of the foregut ectoderm. The diverticulum, which persists throughout life as the **common bile duct**, grows ventrally to invade the mesenchyme of the septum transversum and forms the **hepatic duct, cystic duct, gall bladder, and liver**.

   b)   **Hematopoiesis** begins in the liver during the second month of development and although the developing **bone marrow** (primary ossification centers) begins to produce blood cells during the third month, hematopoietic foci are still present in the liver at birth. At the beginning of the fetal period (nine weeks), the liver comprises 10 percent of the fetal mass.

   c)   Vasculature. The liver receives its arterial blood supply from the **hepatic branch of the celiac artery**.

II. FATE OF THE PRIMITIVE DORSAL MESENTERY

   A. Foregut Mesenteries
      1. Mesoesophagus. Cranial to the septum transversum (future diaphragm), the dorsal mesentery of the foregut is broad and indistinct and at this time, the esophagus with its robust layer of splanchnic mesoderm (including the mesoesophagus) is commonly referred to as the **primitive mediastinum**. The term primitive mediastinum is used to indicate the importance of this area during subsequent partitioning of the coelomic space into separate pericardial, pleural and peritoneal cavities. The splanchnic mesoderm of the mesoesophagus is eventually incorporated into the connective tissues of definitive mediastinal structures.
      2. Mesogastrium **(Greater Omentum).** The **spleen** (lien) develops as a localized accumulation of mesenchymal cells around blood vessels located in the mesogastrium; after the spleen appears, the following subdivisions (peritoneal ligaments) can be recognized in the greater omentum.
         a) **gastrophrenic** - between the stomach and diaphragm above the spleen.
         b) **gastrolienal** - between the stomach and spleen.
         c) **lienorenal** - between the spleen and kidney .
         d) **phrenicocolic** - between the diaphragm and transverse colon.
         e) **gastrocolic** - between stomach and transverse colon.
         f) **omental apron** - covers but is unattached to the remaining abdominal viscera.
      3. Mesoduodenum. The dorsal mesentery is lost during fixation when the duodenum and its associated pancreatic primordia become **retroperitoneal.**

   B. Midgut Mesenteries
      1. Mesentery of the small intestine. The definitive mesentery for the jejunum and ileum is derived almost entirely from the mesentery supporting the cranial limb of the midgut loop during gut rotation; a small segment for the terminal ileum originates from the mesentery of the caudal limb.
      2. Mesoappendix. The mesentery of the appendix represents a small and insignificant derivative of the mesentery for the caudal limb of the midgut loop.
      3. Ascending mesocolon. The mesentery of the ascending colon is normally lost when the ascending colon becomes retroperitoneal by fixation to the right posterior body.
      4. Transverse mesocolon. The primitive transverse mesocolon fuses with the posterior fold or leaf of the greater omentum to form the definitive transverse mesocolon of the adult.

   C. Hindgut Mesenteries
      1. Descending mesocolon. The cephalic portion of the mesentery is normally lost when the descending colon becomes retroperitoneal by fixation to the left posterior body.
      2. Sigmoid mesocolon. Incomplete fusion allows the caudalmost portion of the descending mesocolon to persist as the mesentery of the sigmoid colon.

III. FATE OF THE PRIMITIVE VENTRAL MESENTERY

   A. Lesser Omentum. The lesser omentum is a peritoneal reflection formed as the liver and common bile duct grow beyond the confines of the septum transversum; it is comprised of two peritoneal ligaments.
      1. **Hepatogastric ligament**: between the liver and lesser curvature of the stomach.

2. **Hepatoduodenal ligament:** free edge of the lesser omentum containing the common bile duct, hepatic artery, and portal vein.

B. Hepatic Ligaments. The hepatic ligaments are peritoneal reflections formed when the liver grows beyond the septum transversum and in the process, separates the umbilical vein from the anterior body wall.
   1. **Falciform ligament:** between the anterior body wall and liver; in the adult, its free edge contains the fibrous remnant of the umbilical vein, i.e., **ligamentum teres hepatis**.
   2. **Coronary ligament:** surrounds the bare areas of the liver and diaphragm; the right and left triangular ligaments are part of the coronary ligament.

UROGENITAL SYSTEM

I.   URINARY SYSTEM

A.   Kidney Formation
   1. Three different urinary structures (**pronephros, mesonephros, metanephros**) are formed during the course of early development from **intermediate mesoderm.** Each urinary structure is comprised of a series of **urinary tubules** joined to excretory **urinary ducts** terminating in the cloacal portion of the hindgut.
   2. Pronephric Kidney.
      a) **Pronephroi** develop at cervical levels and are the first to appear. They are very rudimentary in placental mammals.
      b) **Pronephric tubules** degenerate without a trace and are so transitory in humans that they do not appear to develop in all human embryos.
      c) **Pronephric ducts** degenerate cranially but persist at lower levels to serve as excretory ducts for the more caudally located mesonephric tubules, i.e., they become the **mesonephric ducts**.
   3. Mesonephric Kidney.
      a) **Mesonephroi** develop in intermediate mesoderm in thoracic and upper lumbar regions. They function in excretion between the sixth and tenth weeks. In males, a portion of them participate in forming male genital ducts. They regress in females.
      b) **Mesonephric tubules** form excretory units which later degenerate. In males, a few participate in formation of efferent ductules of the male genital system.
      c) **Mesonephric ducts** form from pronephric ducts and drain the mesonephroi into the cloaca. The mesonephric duct persists to form the **ductus epididymis, ductus deferens, seminal vesicle** and **ejaculatory duct** of the genital system.
   4. Metanephric Kidney.
      a) **Metanephroi**, adult functional kidney, develop in intermediate mesoderm in the sacral region in response to inductive signals from the ureteric buds. They begin to function in the tenth week of development.
      b) **Ureteric buds** (metanephric diverticuli) develop as outgrowths from the terminal part of the mesonephric ducts. Ureteric buds grow dorsally and cranially to contact sacral intermediate mesoderm. The distal end of the ureteric bud will be induced by intermediate mesoderm to undergo repeated divisions to form the **pelvis, major calyces, minor calyces** and **collecting ducts**, up to and including the **arched collecting tubules,** for each nephron.

c) Metanephrogenic blastema. Sacral intermediate mesoderm will condense around the distal end of the ureteric bud to form the **metanephrogenic blastema**. The blastema cells induce the ureteric bud to branch. Ureteric bud will induce the metanephrogenic blastema to form the definitive nephrons; **Bowman's capsule, proximal convoluted tubule, loop of Henle, and distal convoluted tubule**.

d) The distal convoluted tubules (metanephrogenic mesoderm) connect with the collecting tubules (ureteric bud) to form a functioning metanephric kidney during the tenth week of development.

5. Ascent of the Adult Kidney.

a) The metanephric kidneys ascend from their sacral location to upper lumbar region.

b) As they ascend they are vascularized by a series of transient vessels off the aorta.

B. Formation of the Urinary Bladder and Urethra

1. Division of Cloaca. The cloaca is divided by the urorectal septum into the ventral urogenital sinus and dorsal rectum.

2. Urogenital Sinus. The urogenital sinus will form from cranial to caudal: a **vesicle part** (bladder attached cranially to allantois), **pelvic part** (males: prostatic urethra; females: membranous urethra), and **phallic part** (males: spongy urethra; females: vestibule of the vagina).

3. Trigone of the Bladder. The terminal portion of the mesonephric duct with its attached ureteric bud is incorporated into the posterior wall of the urogenital sinus (**trigone area**). In males, the incorporation occurs in such a way that the ureteric bud terminates in the upper or urinary portion of the sinus with the mesonephric duct (vas deferens) opening into the lower or genital (prostatic) portion of the sinus. A comparable incorporation occurs in the female, but it is the terminal portion of the fused Müllerian (paramesonephric) ducts which become associated with the genital portion of the sinus.

C. Developmental Defects

1. Ascent of the Kidney.

a) **Ectopic kidney** may be located anywhere along path of ascent.

b) **Abnormal rotational** occurs when the kidney fails to rotate properly during ascent.

c) **Horseshoe kidney** is the result of fusion of the pole of the kidneys and ascent to the level of the inferior mesenteric artery.

d) **Accessory renal arteries** are the result of inadequate regression of caudal vessels during kidney ascent.

2. Duplication of Urinary Tract

a) Duplications of abdominal part of ureter and renal pelvis. These duplications are fairly common and are the result of duplication of distal portion of ureteric bud.

b) Duplications of complete renal system. These duplications are the result of two ureteric buds. The ureter from the cranial kidney is usually ectopic and drains caudal to the ureter from the caudal kidney.

3. Ectopic Ureteric Orifices. In males, an ectopic ureter usually opens into the neck of the bladder or prostatic portion of the urethra or more rarely into ductus deferens, prostatic utricle or seminal vesicle. In females, an ectopic ureter may open into bladder neck, urethra, vagina, or vestibule of the vagina. An ectopic ureter results in **incontinence**

with constant dribbling from the urethra in males and the urethra and/or vagina in females.

4. Renal Agenesis (unilateral or bilateral).
   a) Renal agenesis results when either the ureteric buds fail to form or the buds fail to contact intermediate mesoderm.
   b) Bilateral renal agenesis is associated with **oligohydramnios** because no urine is secreted into amniotic cavity. Infants with bilateral renal agenesis die shortly after birth.

II. REPRODUCTIVE SYSTEM

A. Determination of Genetic Sex
   1. Genetic sex (XX or XY) is determined at the time of fertilization.
   2. Gonadal type (ovary or testis) are determined by the genetic sex.
   3. Genital duct and external genitalia development is **not dependent on genetic sex**. Development is dependent on presence or absence of a testis. Differentiation of genital ducts and external genitalia in females **does not** require hormonal stimulation and will occur in the absence of ovaries.

B. Development of Gonads
   1. Indifferent stage. Initial stages of gonadal development are similar in males and females.
      a) **Gonadal ridges** forms on the medial side of the mesonephroi, produced by proliferation of underlying mesenchyme.
      b) **Primary sex cords** form as mesothelium grows into underlying mesenchyme.
      c) **Primordial germ cells** are the progenitor cells of the gametes. They arise from endodermal cells of the yolk sac and migrate through the dorsal mesentery to the gonadal ridges.
      d) Gonadal development dependent on the chromosomes present. The Y chromosome has a strong, testis-determining effect on the indifferent gonad. The absence of a Y chromosome results in the formation of an ovary.
   2. Testes.
      a) Seminiferous cords and tubules form from primary sex cords.
      b) Primordial germ cells that entered primary sex cords form **spermatogonia**.
      c) **Spermatogenesis** does not begin until puberty and continues throughout life.
      d) Sertoli cells of the embryonic seminiferous tubules produce **Müllerian inhibiting substance**, a peptide of member of the transforming growth factor family.
      e) Interstitial cells of Leydig produce **testosterone**, an androgenic steroid.
   3. Ovaries.
      a) **Cortical cords** form from a secondary invasion of mesothelium into mesenchyme. The primary sex cords are lost.
      b) Primordial germ cells invade the cortical cords and form **oogonia**.
      c) Oogonia begin the process of **oogenesis** during development. They enter the first meiotic division to become primary oocytes and become arrested in prophase I.
      d) Primordial follicles form when the cortical cords disrupt, forming a single layer of follicle cells around the primary oocyte.
      e) All gametes have developed into primary oocytes by the time of birth and remain at this stage until stimulated to develop further during the reproductive cycle.

C. Development of Genital Ducts
1. Indifferent stage. During the indifferent stage of development every embryo possesses the primordia for both **male** and **female genital ducts**, i.e., **mesonephric and Müllerian (paramesonephric) ducts**.
2. Differentiation of genital duct primordium. The endocrine portion of the embryonic testis becomes functional during the seventh week of development and produces **testosterone and Müllerian inhibiting substance** which cause the indifferent genital ducts to differentiate into those of a male.
3. Fate of the genital duct primordium
   a) Males
      (1) Testosterone. The presence of testosterone stimulates the growth and development of the male primordia (mesonephric tubules and ducts) and assures development of the appropriate genital ducts for a testis (efferent ductules, ductus epididymis, ductus deferens, seminal vesicles, ejaculatory ducts).
      (2) Müllerian Inhibiting Substance. The presence of Müllerian Inhibiting Substance causes the female primordia (Müllerian or paramesonephric ducts) to degenerate. This is why remnants of the female genital ducts are rare in normal males.
   b) Females
      (1) Absence of Müllerian Inhibiting Substance. The absence of Müllerian Inhibiting Substance allows the female primordia (paramesonephric ducts) to persist and differentiate as the definitive genital ducts. The upper ends of the ducts persist as the **uterine tubes**; the lower ends fuse to form the **uterus** and to induce vaginal development.
      (2) Absence of Testosterone. The absence of testosterone causes growth and development of the male primordia to cease but not degenerate; this is why remnants of the male genital ducts (**Gartner's cysts and ducts**) are commonly seen in the broad ligament and vaginal wall of normal females.
4. Developmental Defects. These occur more frequently in males than in females because differentiation of the male reproductive system requires a functional gonad and hormonal stimulation while that of the female will differentiate normally without an ovary or hormonal stimulation. As a consequence of the hormonal dependency in males, any defect in either the biosynthetic pathway of testosterone or in the expression of its masculinizing effects will be reflected by partial or complete feminization, i.e., incomplete virilization of the genital ducts and/or external genitalia.

D. Formation of the Vagina and Prostatic Utricle
1. Vagina
   a) **Sinovaginal plate** represents the primordium of the vagina. It develops as a thickening of endoderm of urogenital sinus and is induced by the paramesonephric ducts.
   b) The sinovaginal plate expands to form the vagina which later canalizes. The hymen marks the junctional area between sinovaginal plate and urogenital sinus.
2. Prostatic Utricle
   a) The prostatic utricle is considered to be the homologue of the female vagina. The lower ends of the paramesonephric ducts must persist in male development long enough to induce the formation of a sinovaginal plate for formation of prostatic utricle.

E. Development of the External Genitalia
1. Indifferent stage. The indifferent stage (pudendal primordium) is composed of a **genital tubercle (phallus)**, **urogenital folds** surrounding a urogenital groove which opens into the differentiating urogenital sinus, and **labioscrotal swelling** surrounding the urogenital folds.
2. Differentiation of the pudendal primordium. In the pudendal primordium, the appropriate masculinizing response in androgen sensitive cells is assured by the appearance of an enzyme (5-alpha reductase) which converts testosterone to **dihydrotestosterone**; the latter androgenic steroid is approximately fifty times more potent than testosterone. Increased potency of dihydrotestosterone compensates for the dilutional effects resulting from a small embryonic testis and a relatively large circulating blood volume (embryonic plus placental).
3. Fate of the pudendal primordium.
   a) **Males.** Androgenic stimulation of the indifferent pudendal primordium produces masculinization by causing greater development of the genital tubercle and by promoting fusion of the labioscrotal swellings and urethral folds, i.e., **penis**, **scrotum** and **penile** urethra formation.
   b) **Females.** The absence of androgenic stimulation results in less development of the genital tubercle and in failure of the labioscrotal swellings and urethral folds to fuse, i.e., **clitoris**, **labia majora** and **labia minor** formation.
4. Developmental Defects.
   a) Masculinization of the pudendal primordium requires both testosterone (as a prohormone) and dihydrotestosterone. As a consequence, any defect in the biosynthesis of either, e.g., deficiency of the reductase, or in the expression of their masculinizing effects (lack of receptors) will result in partial or complete feminization of the pudendal primordium, i.e., **ambiguous genitalia**.
   b) **Male pseudohermaphrodites** are classic examples of individuals with ambiguous genitalia due to incomplete masculinization (partial feminization). The range of ambiguity varies from forms resembling perineal hypospadias to those requiring cytogenetic analysis to determine genetic sex.
   c) **Testicular feminization syndrome.** These individuals are genetic males, with normal or above normal androgen levels and complete feminization of external genitalia. Masculinization fails to occur because the peripheral tissues lack androgen receptors, i.e., the defect occurs at the expression level.
   d) **Female pseudohermaphrodites** are relatively rare but may be produced when female embryos are exposed to androgenic or potentially androgenic substances. The source of the androgens may be endogenous or exogenous. The best example of female pseudohermaphrodites originating from endogenous androgens is seen in individuals with **adrenogenital syndrome** (congenital virilizing adrenal hyperplasia). In this syndrome, endogenous androgenic substances originate from the adrenal cortex resulting masculinization. Female pseudohermaphrodites may also occur when exogenous androgenic substances, e.g., progesterone, are administered to the mother during pregnancy.

DEVELOPMENT OF THE HEART

I. FORMATION OF THE PRIMARY HEART TUBE

A. Lateral Endocardial Tubes.

1. The heart, like all other blood vessels, appears first as a simple **endothelial tube**. (In this case, "endo" refers to the epithelium lining all parts of the cardiovascular system **not** endoderm; all vascular elements are derived from mesoderm). In the definitive heart, the endothelium forming the 'heart tubes' persists as the epithelial layer of the **endocardium**; the cardiac muscle of the **myocardium** develops from splanchnic mesoderm.

2. Initially a pair of **lateral endocardial tubes** develop in the splanchnic mesoderm in the **cardiogenic region**.

B. Positioning of the Heart

1. **Heading folding** moves the cardiogenic region with developing heart ventrally and caudally. The heart is suspended in the pericardial cavity by a dorsal mesentery (**mesocardium**). A short time later, degeneration of the central portion of the mesocardium leaves the heart attached only at its cranial (arterial) and caudal (venous or septal) ends. The intermediate area now devoid of mesentery persists as the **transverse sinus** of the adult pericardial cavity.

2. Lateral folding brings the paired lateral endocardial tubes to form the single **primary heart tube**.

II. DIFFERENTIATION OF THE PRIMARY HEART TUBE

A. Four Primitive Cardiac Chambers. The appearance of regional dilations or expansions in the tubular heart subdivides the heart into four primitive cardiac chambers.

1. **Sinus venosus** or most caudal region of the heart which receives all of the blood being returned to the heart from the three interdependent circuits.

a) **Common cardinal veins** returning blood from the somatopleure (body wall) of the embryo.

b) **Vitelline veins** returning blood from the splanchnopleure (gut and yolk sac).

c) **Umbilical (allantoic) veins** returning blood from the placenta.

2. Primitive atrium

3. Primitive ventricle

4. **Bulbus cordis** and **truncus arteriosus** which form the most cranial portion of the heart. The truncus arteriosus represents the fused portions of the primitive ventral aortae; the unfused portions are continued cranially and dorsally as the first pair of aortic arches.

B. Heart Tube Folding

1. The heart tube elongates more rapidly than the surrounding pericardial cavity resulting in folding. The **bulbus cordis** is moved ventrocaudally and to the right while the **primitive ventricle** is moved to the left and the **primitive atrium** craniodorsally.

2. Normal folding to the left is called **sinistral** folding. Abnormal folding to the right is called **dextrocardia**.

C. Fate of the Four Primitive Cardiac Chambers

1. The **primitive atrium** becomes divided into right and left atria and the sinus venosus becomes incorporated into the definitive right atrium.

2. The **primitive ventricle** forms all of the left ventricle of the adult heart except the outflow area of the aorta.

3. The **bulbus cordis** forms all of the right ventricle including the pulmonary outflow area and the aortic outflow area of the left ventricle.

4. The **truncus arteriosus** (ventral aorta) becomes partitioned to form the origin for the aorta and for the pulmonary artery.

D. Fate of the Sinus Venosus. Initially, the opening between the sinus venosus and the primitive atrium is located in the midline but as venous blood is shunted to the right side of the heart, the right extremity of the sinus venosus (right horn) enlarges moving the sinoatrial opening to the right. Later, the right horn of the sinus venosus becomes incorporated into the primitive atrium to form the **smooth portion** of the definitive right atrium. The left horn of the sinus venosus remains small because venous return is being shifted to the right side of the heart; it persists in the adult as the **coronary sinus** and receives only venous blood from the heart itself.

E. Fate of the Primitive Pulmonary Veins. The terminal portion of the primitive pulmonary veins are incorporated into the left half of the primitive atrium to form the **smooth portion** of the left atrium. Resorption of the primitive pulmonary vessels extends distally to the appropriate bifurcation level to produce the four definitive pulmonary veins entering the adult heart.

III. DEVELOPMENT OF THE FOUR CHAMBERED HEART
As the tubular heart changes shape and position, a series of septa develop which convert it into four chambers. All of the changes occurring in the developing cardiovascular system, heart and vessels, are taking place simultaneously. In addition all of these changes occur without interrupting the continuous pumping action of the heart.

A. Division of Common Atrioventricular Opening.
1. Formation of endocardial cushions. Localized proliferations of mesenchyme within the anterior and posterior walls of the common atrioventricular opening result in the appearance of two elevations, the **endocardial cushions**. Subsequent growth and fusion of the dorsal and ventral endocardial cushions divides the common atrioventricular opening into definitive **right** (tricuspid) and **left** (mitral) **atrioventricular canals**. The fused endocardial cushions participate in separation of the definitive atria and ventricles by contributing to the formation of the interatrial (septum primum) and interventricular (membranous portion) septa.
2. Developmental defects. A **common atrioventricular canal** is caused by failure of the endocardial cushions to fuse; this is one of the common cardiac defects associated with **Down's syndrome**. The same fusion failure of the dorsal and ventral endocardial cushions would also disrupt their subsequent fusion with the septum primum and contribute to atrial septal defects. Displacement to the right or left would produce tricuspid or mitral stenosis with dilatation of the corresponding canal on the opposite side. Displacement to either side would also interfere with its contributions to the membranous septum and contribute to high interventricular septal defects.

B. Formation of the Interatrial Septum. The formation of two septa (**primum and secundum**) is required to effect complete separation of the single primitive atrium into the definitive right and left atria of the adult heart. Although separated, **right-to-left shunting of blood** through the foramen ovale persists until birth.
1. Septum primum. The **septum primum** forms as a crescentic-shaped partition projecting inferiorly forming the **foramen primum**, an ostium between the developing right and left atria. The septum primum grows inferiorly to fuse with the endocardial cushions and obliterate the foramen primum. The upper portion of the septum primum degenerates to produce a new ostium, **foramen secundum**.

2. Septum secundum. The septum secundum is also crescentic in shape and develops immediately to the right of septum primum. Septum secundum grows dorsally and inferiorly but does **not** fuse with the endocardial cushions. The opening remaining is the **foramen ovale**. The persistent, lower portion of septum primum functions as the **valve of the foramen ovale**.

3. Developmental defects. Maldevelopment of the septum primum or septum secundum will result in **atrial septal defects**. Most involve defective formation of the foramen ovale.

C. Formation of the Definitive Ventricles
1. Fate of primitive ventricle and bulbis cordis.
   a) The **primitive ventricle** forms all of the definitive left ventricle **except** the aortic outlet.
   b) The **bulbus cordis** forms all of the definitive right ventricle **and** the aortic outlet of the left ventricle.
2. Interventricular septum.
   a) The broad communication between the primitive ventricle and bulbis cordis is partially closed by the **inferior bulboventricular ridge** which forms the **muscular portion of the interventricular septum**.
   b) The formation of the **membranous part of the interventricular septum** completes separation of definitive right and left ventricles. The formation of the membranous septum involves the proliferation and fusion of mesenchyme from the aorticopulmonary septum and endocardial cushions.
3. Developmental defects. Most **ventricular septal defects** result from displacement of either the aorticopulmonary septum or endocardial cushions which interfere with their fusion and formation of a membranous interventricular septum.

D. Formation of the Aortic and Pulmonary Trunks
1. Fate of the truncus arteriosus. The truncus arteriosus is divided into aortic and pulmonary trunks by formation of the **aorticopulmonary** or **spiral septum**. The aorticopulmonary septum develops from two spiraling ridges of mesenchyme which project into the lumen of the truncus arteriosus; subsequent growth and fusion of the spiral ridges produce the separate lumina for the **aortic** and **pulmonary trunks**.
2. Developmental defects. Complete absence of the aorticopulmonary septum results in persistent truncus arteriosus; incomplete fusion of the ridges results in aortico-pulmonary communications; incomplete spiraling results in transposition of the great vessels; displacement of the ridges with unequal division of the truncus results in pulmonary or aortic stenosis with dilatation of the opposite vessel. Displacement of the lower ends of the spiral ridges interferes with formation of the membranous septum and results in high interventricular septal defect, i.e., persistent interventricular foramen. The anatomical defects in the **tetralogy of Fallot** (pulmonary stenosis, ventricular septal defect, overriding aorta) are thought to arise in this way (the right ventricular hypertrophy is a physiological response to the pulmonary stenosis).

DEVELOPMENT OF THE VASCULATURE

I. VASCULAR DEVELOPMENT

    A. Blood Vessel Formation
       1. Blood vessels develop from extraembryonic and embryonic mesoderm.
       2. Extraembryonic mesoderm. Blood vessels form in extraembryonic mesoderm of yolk sac, connecting stalk and chorion from **blood islands**. **Endothelial cells** and **hemoblasts** (primitive blood cells) form from blood islands. Form interconnecting vascular network.
       3. Embryonic mesoderm. Blood vessels form in splanchnic mesoderm in embryo from **angioblasts**. Angioblasts form endothelial cells and interconnecting vascular network by process of **vasculogenesis**.

    B. Hematopoiesis. Primitive blood cells form from hemoblasts in blood islands in yolk sac. Later, blood cells form in the embryo; initially in the liver and later in bone marrow.

II. DEVELOPMENT OF THE ARTERIAL SYSTEM

    A. Aortic Arches
    Each **branchial arch** in the branchial apparatus contains an aortic arch which connects the ventral truncus arteriosus and dorsal aortae. In lower vertebrates, the aortic arches form the functional gills. Although the respiratory function for most of the branchial arches has been lost, **six aortic arches** still appear during human development. Three of these persist (three, four, six) and make definitive contributions to the adult vasculature but only the sixth has retained a respiratory function. Their appearance in the human embryo is often transitory and all six are not present at the same time.
       1. Aortic arch 1. It is the first to appear and is formed as a result of head fold formation; **it disappears soon after development**. The mandibular and maxillary areas of the adult are supplied by new vessels arising from the external carotid
       2. Aortic arch 2. Persists only during the early stages of development.
       3. Aortic arch 3. Persists in the adult to form the **origin of the internal carotid artery**. The dorsal aorta between aortic arches 3 and 4 disappears and the extension of the internal carotid into the cranial area is an active growth process of the dorsal aorta, to supply the rapidly enlarging brain.
       4. Aortic arch 4. Persists on the left side as the **arch of the aorta**. On the right side it contributes to the **brachiocephalic artery** and the proximal portion of the **right subclavian artery**.
       5. Aortic arch 5. Extremely transitory and does not appear to develop in all human embryos. The developmental variability appears to be related to the rudimentary condition of the fifth branchial arch.
       6. Aortic arch 6. Well developed and forms the origin or proximal portion of the **pulmonary arteries**; the distal portions of the pulmonary arteries arise directly from the sixth arch and accompany the developing lung buds into the pleural cavities. On the right side the original connection of the sixth arch to the dorsal aorta is lost and the **right recurrent laryngeal nerve** (the nerve for the branchiomeric musculature developing from the sixth arch) slips cranially to loop around the right subclavian artery (fourth arch). On the left side, the connection between sixth arch and dorsal aorta is

retained as the **ductus arteriosus** and as a consequence the **left recurrent laryngeal nerve** descends into the thorax to 'recur' around the ligamentum arteriosum.

    7.   Developmental defects. Coarctation of the aorta.

        a)   **Postductal coarctation** is the most common form and is usually located immediately below the origin of the left subclavian artery. Postductal coarctation probably has its origin during fetal life but, because of better collateral circulation, this form of coarctation may go undetected for years. It is for this reason that it is sometimes referred to as the **adult form** of coarctation.

        b)   **Preductal coarctations** are located immediately above the ductus arteriosus; the ductus itself is almost invariable patent. The length of the aortic segment involved may be extensive and its location precludes development of collateral circulation. Without surgical intervention, the prognosis is poor. Since the condition is detected at or shortly after birth, the preductal narrowing is frequently referred to as the **infantile form of coarctation**.

  B.   Branches of Dorsal Aorta

    1.   The dorsal aorta develops **ventral, lateral and dorsolateral branches**.

    2.   Ventral branches. Vitelline arteries supply blood to the yolk sac. Later they condense to form arteries which supply the gut tube and its derivatives.

        a)   Thoracic foregut receives about four vitelline arteries.

        b)   Abdominal foregut supplied by **celiac artery**.

        c)   Midgut supplied by **superior mesenteric artery**.

        d)   Hindgut supplied by **inferior mesenteric artery**.

III.   DEVELOPMENT OF THE VENOUS SYSTEM

  A.   Venous Drainage of Early Embryo

    1.   The venous drainage pattern of the early human embryo consists of three venous systems.**Umbilical (allantoic) veins**. Return blood from the placenta carrying oxygenated blood.

    2.   **Vitelline (omphalomesenteric) veins.** Return blood from the splanchnopleure, initially the yolk sac and later the gut tube and its derivatives.

    3.   **Cardinal veins**. Composed of anterior, posterior and common. Return blood from the somatopleure of the embryo. The **anterior cardinal veins** drain the cranial levels of the embryo, above the septum transversum. The **posterior cardinal veins** drain the caudal levels of the embryo, below the septum transversum. The terminal portions of anterior and posterior unite to form **common cardinal veins** (ducts of Cuvier). All of these vessels pass through the septum transversum before terminating in the sinus venosus.

  B.   Fate of the Umbilical Veins
The **right umbilical vein** disappears during early development. The **left umbilical vein** remains functional. In the septum transversum the developing liver surrounds the terminal part of the left umbilical vein forming the **ductus venosus**.

  C.   Fate of the Vitelline Veins

    1.   The vitelline veins form the **hepatic portal system**.

    2.   The **hepatic portal vein** is derive from the extrahepatic portions of both vitelline veins.

    3.   The intrahepatic portions of the vitelline veins are invaded by hepatic cords but persist as **hepatic sinusoids, central, sublobular** and **hepatic veins** of the adult liver.

D. Fate of the Anterior, Posterior and Common Cardinal Veins
1. Anterior cardinal veins persist as the **internal jugular veins**. The terminal portion of the right anterior cardinal vein forms the **superior vena cava**. A new vessel (**left brachiocephalic**) forms across the midline to shunt blood from left anterior cardinal to the right. When the left brachiocephalic vein fails to form, the terminal portion of the left anterior cardinal vein remains functional and forms a left superior vena cava terminating via the oblique vein in the coronary sinus.
2. Posterior cardinal veins regress and are replaced by a new somatopleuric vessel, the **inferior vena cava**, which originates from capillary plexi above and below the posterior cardinals, the supracardinal and subcardinal plexi.

E. Formation of the Inferior Vena Cava
1. The upper portion of the inferior vena cava from the diaphragm to and including the renal veins, is derived from the **subcardinal midline anastomoses**. This is why the upper portion of the inferior vena cava is anterior or "sub" to the aorta.
2. The lower portion of the inferior vena cava, below the renal veins, is derived from the **supracardinal midline anastomoses**. This is why the lower portion of the inferior vena cava is posterior or "supra" to the aorta.

IV. FETAL CIRCULATION

A. Pathway of Oxygenated Blood
1. Oxygenated blood leaves the placenta via the **left umbilical vein** and is shunted through the liver via the **ductus venosus** to enter the **inferior vena cava**.
2. Within the inferior vena cava the umbilical blood is mixed with smaller amounts of venous blood from the caudal levels of the body.
3. Oxygenated blood in the inferior vena cava enters the **right atrium** and is diverted through the **foramen ovale** into the **left atrium**. Blood passes through the **left atrioventricular (mitral) canal** and leaves the **left ventricle** via the **ascending aorta**.
4. From the **arch of the aorta**, the oxygenated blood is distributed via the **brachiocephalic trunk** and the **left common cardinal and subclavian arteries** to the upper portion of the body. The intracardiac shunt via the foramen ovale assures that the head and especially the developing brain receives the most highly oxygenated blood possible.

B. Pathway of Deoxygenated blood
1. Deoxygenate blood from the upper half of the body enters the **right atrium** from the **superior vena cava** and passes directly through the **right atrioventricular (tricuspid) canal** to enter the **right ventricle** and **pulmonary trunk**.
2. A small portion of the deoxygenated blood enters the pulmonary arteries to perfuse the developing lungs but the major portion is shunted via the **ductus arteriosus** into the **thoracic aorta** below the origin of the left subclavian artery. Oxygenated and deoxygenated blood mix at this point.
3. Deoxygenated blood is carried via the **umbilical arteries** to the **placenta**.

V. CIRCULATORY CHANGES AT BIRTH

A. Fate of the Foramen Ovale
1. At birth, the right to left shunt provided by the foramen ovale ceases to function when changes in atrial pressure bring the overlapping portions of the **septum primum** (valve of the foramen ovale) and the **septum secundum** into apposition.

2. **Changes in atrial pressure** are the results of an abrupt decrease in right atrial pressure with sudden closure of umbilical vein and increase in left atrial pressure associated with increased perfusion and venous return of pulmonary circulation.
3. Failure of the foramen ovale to close anatomically (fusion of the septum primum and secundum) is relatively common, i.e. **probe patent foramen ovale**, but is insignificant physiologically.
4. In the adult heart, the location and boundaries of the foramen ovale are indicated by the **limbus fossae ovalis**. The **floor of the fossa ovalis** is formed by the septum primum.

B. Fate of the Ductus Arteriosus
1. The ductus arteriosus begins to close almost immediately after birth and may be closed functionally within a few hours. Complete obliteration of the lumen may require several weeks. The ductus arteriosus persists throughout life as the **ligamentum arteriosum**.
2. **Patent ductus arteriosus (PDA)**. If the ductus arteriosus remains patent, aortic blood is shunted into the pulmonary artery. Premature infants usually have PDA for the first 24 hrs of postnatal life. PDA may result from other malformations such as preductal coarctation of the aorta, pulmonary stenosis or atresia, or transposition of the great vessels. PDA may result from **maternal rubella infection** during early pregnancy.

C. Fate of the Umbilical Vein and Ductus Venosus
After interruption of the placental circulation, the extra- and intrahepatic portions of the umbilical vein become reduced to fibrotic cords which persist in the adult as the **ligamentum teres hepatis** and **ligamentum venosum** respectively

D. Fate of the Umbilical Arteries
In the adult, the proximal portions of the umbilical arteries persist as the umbilical branches of the internal iliacs; the distal portions which accompany the intraembryonic portion of the allantois (urachus) are obliterated but persist as the **medial umbilical ligaments**; the urachus itself persists as the **median umbilical ligament**.

GROSS ANATOMY

BASIC CONCEPTS OF THE NERVOUS SYSTEM AND IMAGING

I.   Overview of nervous system: Functional cellular unit is a neuron. Extensions from the cell body of the neuron, called dendrites, receive information as electrical impulses or action potentials; an extension from the nerve cell body called an axon transmits information away from the nerve cell body. Interneuronal communication occurs via the synapse, a small space between neurons across which action potentials are transferred by neurochemicals to the adjacent neuron. Efferent (motor) neurons carry information away from the central nervous system; afferent (sensory) neurons carry information toward the central nervous system. Efferent or afferent neurons supplying the body wall and limbs are called somatic, efferent or afferent neurons supplying organs and blood vessels are called visceral.

A.   Central nervous system: Consists of the brain and spinal cord. Membranes called meninges cover the brain and spinal cord. Three meninges: outer tough layer called dura mater, intermediate delicate layer called arachnoid mater, and innermost thin layer inseparable from brain and spinal cord, called pia mater. Pia mater follows contours of sulci and gyri in the central nervous system. In the spinal cord, the denticulate ligament and filum terminale are derived from pia mater. Denticulate=lateral tooth like extensions of pia, between dorsal and ventral roots of spinal nerves, that attach to dura and limit twisting of spinal cord within the dural sac. Filum terminale=delicate, vascular, membranous caudal continuation of spinal cord. No nerve cells in filum terminale. Arachnoid and dura lie across sulci and gyri. Deep to the arachnoid, the subarachnoid space contains cerebrospinal fluid. In the spinal cord, there is a potential subdural space which normally contains only a film of fluid. In the cranium, the dura consists of two tough layers: outer layer (periosteal dura) functions as periosteum for the skull and the inner layer (meningeal dura) dips between and separates and supports lobes and hemispheres of the brain. The meningeal or true dura also forms pathways for the various dural venous sinuses.

B.   Peripheral nervous system: consists of 31 spinal nerves (8 cervical, 12 thoracic, 5 lumbar, 5 sacral, 1 coccygeal) and 12 cranial nerves (numbered 1 through 12 and also named).
1.   Typical spinal nerve.
a)   Morphology. Dorsal roots (sensory) and ventral roots (motor) attach to the dorsal and ventral aspects of the spinal cord. Laterally, near the intervertebral foramen, the dorsal (posterior) root has a swelling, the dorsal root ganglion, which is formed by the nerve cell bodies of afferent components. At the entrance to the intervertebral foramen, dorsal and ventral (anterior) roots converge to form the mixed spinal nerve. Lateral to the intervertebral canal, the mixed (motor and sensory) spinal nerve bifurcates into 2 main mixed branches, the dorsal and ventral primary rami. Dorsal rami bifurcate a second time forming medial and lateral branches. Dorsal rami innervate dorsal axial musculature and provide sensation to the dorsum. The ventral ramus innervates anterolateral musculature and supplies sensation to the anterolateral skin. Each ventral ramus gives off a lateral cutaneous branch at the mid-axillary line which divides into anterior and posterior branches. The ventral ramus continues anteriorly to terminate as an anterior cutaneous branch. Dermatome=area of skin supplied by a spinal nerve and is important in clinical diagnosis.

b) Functional components in fibers of typical spinal nerve:
   GSE= General somatic efferent, motor to voluntary muscle
   GSA= General somatic afferent, sensory from somatic receptors
   GVE= General visceral efferent, autonomic to involuntary muscle, cardiac muscle and glands
   GVA= General visceral afferent, sensory from visceral receptors
c) Reflex arc: when a sensory receptor is stimulated, an impulse travels over an afferent process to reach the spinal cord and synapse on an efferent neuron which mediates an impulse response to an effector organ.

2. Autonomic nervous system (ANS). This may also be referred to as the GVE system. This portion of the central and peripheral nervous system provides innervation to smooth muscle, cardiac muscle and glands over which we have little to no volitional control. The number 2 seems to be recurrent in the ANS: There are 2 major divisions, each major division has two anatomical outflows from the CNS, and each division utilizes 2 efferent neurons (preganglionic and postganglionic) to reach effector structures. GVA components (sensory) usually accompany GVE components in the peripheral nervous system.

a) Sympathetic=thoracolumbar portion. Preganglionic nerve cell body resides in intermediolateral cell column in thoracic (T_{1-12}) and lumbar (L_1 to L_2 or L_3) segments of spinal cord. Preganglionic process leaves CNS as component of ventral root of spinal nerve. Preganglionic process leaves ventral root as component of white ramus communicans. Postganglionic nerve cell bodies reside in paravertebral and prevertebral ganglia. Preganglionic processes may ascend in the sympathetic trunk, descend in the sympathetic trunk, or synapse in a sympathetic trunk (paravertebral) ganglion. Postganglionic processes connecting paravertebral ganglia to spinal nerves constitute grey rami communicans. To recap: white ramus communicans transmit preganglionic nerve cell processes, grey ramus communicans transmit post ganglionic nerve cell processes. The preganglionic process at some levels traverses the paravertebral ganglion without synapsing. In such cases the preganglionic process exits the paravertebral ganglion as a component of a sympathetic splanchnic (thoracic, lumbar, sacral) nerve. Preganglionic components in sympathetic splanchnic nerves synapse on postganglionic nerve cell bodies in a prevertebral ganglion. Processes of post ganglionic nerve cell bodies located in prevertebral ganglia usually follow arteries to reach their effector organs (cardiac muscle, smooth muscle, glands). To recap: Two efferent neurons, a preganglionic and a postganglionic neuron are required to go from the spinal cord to the effector structure. Sensory components (GVA), primarily concerned with pain from viscera, accompany GVE sympathetic components. Sympathetic nerves distribute to involuntary muscle and sweat glands located in the trunk and upper and lower extremities.

b) Parasympathetic=Craniosacral portion. Preganglionic nerve cell body resides in nuclei in brain for cranial outflow and intermediolateral column of sacral spinal cord for sacral outflow. In head, preganglionic components leave brain as components of cranial nerves III, VII, IX, and X and synapse on postganglionic nerve cell bodies in named ganglia which are on or near peripheral effector structures. Since the ganglia containing the postganglionic nerve cell bodies are close to the effector structures, the postganglionic process is usually short. In the sacral cord, preganglionics leave spinal cord as components of pelvic splanchnic nerves and again synapse on postganglionic neurons which are components of ganglia close to or on pelvic viscera. Once again, the post-ganglionic process is relatively short. Parasympathetic nerves only distribute to the head and trunk. They do not enter the extremities. Autonomic regulation in the extremities is accomplished by increasing and decreasing sympathetic stimulation.

II. Diagnostic imaging.

A. Densities of radiographs: Radiographs of body tissues exhibit 4 basis densities. In order of increasing radiodensity these are air, fat, water, and bone.

B. Radiographs: Whenever possible radiograph body parts from 2 projections, usually at right angles.

C. Viewing radiographs: View radiographs of a patient as though the patient was facing you; the patient's left side is on your right and the patient's right side is on your left.
   1. Image sharpness and magnification. For sharp images, the object should be close to the film. Close to the film=less magnification. Movement of object away from film=increased magnification and usually less well defined image.
   2. PA vs AP radiographs. Radiographs described in terms of direction beam passes through patient. PA=X-ray tube posteriorly, film anteriorly; similarly, AP=X-ray tube anteriorly, film posteriorly.
   3. Contrast agents: used to examine body parts that do not have inherent density or contrast differences. Example: barium sulfate.
   4. Fluoroscopy. Provides for dynamic image of body part. Fluorescent screen substituted for X-ray film and image is viewed on TV monitor. X-ray tube is usually below table and fluorescent screen is usually above table.
   5. Ultrasound. Non-invasive imaging technique using sonic energy high above human hearing range. Not considered detrimental, therefore used routinely to evaluate fetal development. A transducer is placed against area to be examined; ultrasound waves are transmitted into the body, structures with different acoustical characteristics reflect sound back and the reflected sound is processed electronically to produce an image.
   6. Computed tomography: Measures X-ray absorption of various tissues by a paired X-ray tube and detector system. A computer reconstructs the image based on density measurements.
   7. Nuclear imaging: A radioactive substance is injected into the body and subsequently taken up by certain organs in the body, and the uptake pattern is visualized electronically.
   8. Nuclear magnetic resonance (NMR) and magnetic resonance imaging (MRI). Image production resulting from absorption or emission of electromagnetic energy by nuclei, after excitation, in a magnetic field.

BACK AND UPPER LIMB

I. Vertebral column. The segmentally arranged vertebrae are the result of segmentation of paraxial mesoderm into somites in the embryo (see Muscle skeletal system in Embryology section). The central longitudinal axis which serves to support the trunk and protect the spinal cord. Composed of approximately 33 discrete bones or vertebrae separated by fibrocartilaginous intervertebral disks and united by joints and ligaments. This arrangement provides for flexibility and strength. There are 7 cervical, 12 thoracic, 5 lumbar, 5 sacral (fused to form sacrum) and 3-5 (usually 4) coccygeal vertebrae. Vertebrae tend to increase in mass from the cervical region caudally to accommodate for their increased weight bearing function.

A.   Surface anatomy. Median furrow lies over tips of spinous processes and allows palpation of these processes except in cervical region, where the thickened supraspinal ligament, the ligamentum nuchae, prevents palpation. However, upon flexion, the spine of C7 is usually distinct and is referred to as vertebrae prominens. Superior angle of scapula is at level of T1 or T2 spines and the inferior angle is at level of T7 or T8 spines. The highest point of the iliac crest is at the space between L3 and L4.

B.   Generalized plan. Each vertebrae has body anteriorly, which is weight bearing, which is connected to bodies of adjacent vertebrae via intervertebral disks. Posteriorly, the vertebral arch protects the spinal cord and is not weight bearing. The vertebral (neural) arch circumscribes the vertebral foramen. Vertebral foramina collectively form the vertebral canal which runs the full length of the vertebral column.
1.   Vertebral arch: formed by 2 pedicles and 2 laminae from which arise 2 superior articular processes and 2 inferior articular processes, 2 transverse processes and l spinous process. Superior and inferior articular processes are located at junctions of the pedicles and laminae. One spinous process is located at posterior midline at junction of two laminae. Transverse processes project on either side at junction of laminae and pedicles. Muscles and ligaments attach to transverse and spinous processes.
2.   Intervertebral foramina: bounded by superior and inferior notches of adjacent pedicles and intervertebral disks. Transmits spinal nerve and accompanying vessels.

C.   Regional characteristics:
1.   Atlas (C1): highly specialized, ringlike, thin anterior and posterior arches and 2 lateral masses. No body or spinous process. Large horizontally oriented superior articular processes in the lateral masses have large facets which articulate with condyles of occipital bone.
2.   Axis (C2): has superiorly directed odontoid process (dens) which forms pivot upon which atlas rotates. Dens probably represents "lost" body of C1.
3.   C3 to C6 cervical vertebrae: have delicate bodies, bifid spinous processes, foramina in transverse processes and enlarged triangular vertebral foramen to accommodate cervical enlargement of spinal cord.
4.   C7: has prominent spine (vertebrae prominens).
5.   General feature of cervical vertebrae: Articular processes tend to be oriented horizontally.
6.   Thoracic: Round vertebral foramen, facets on side of body for articulation with heads of ribs and on transverse processes for articulation with tubercle of ribs. Long, inferiorly directed spinous processes overlap. Articular processes oriented in frontal plane in the superior portion of the thoracic vertebral column, inferiorly, they tend to become sagittal in orientation.
7.   Lumbar: massive bodies, short stout pedicles, long transverse processes, short, wide spinous processes. Vertebral foramina triangular to accommodate lumbar enlargement of spinal cord. Articular processes oriented predominately in sagittal plane.
8.   Sacrum: formed by fusion of 5 sacral vertebrae. Anterior surface concave and foramina transmit ventral primary rami of the upper 4 sacral spinal nerves. Posterior surface has 4 foramina which transmit dorsal primary rami of upper sacral spinal nerves. Vertebral canal continues into the sacrum. Laminae of the fifth segment and sometimes fourth,

fail to meet and thus produce the sacral hiatus, an inferior entrance to the vertebral canal. The sacral hiatus may be used to administer epidural anesthesia.

9. Coccyx: consists of 3-5 (usually 4) rudimentary vertebrae at caudal end of vertebral column.

D. Curvatures. Primary curvatures form during fetal period, persist after birth, and include the thoracic and sacral curvatures. These curvatures are concave anteriorly. Secondary curvatures develop after birth, and include the cervical and lumbar curvatures. These develop in response to lifting the head and walking, respectively. Kyphosis=exaggeration of thoracic curvature, lordosis=exaggeration of lumbar curvature, scoliosis is a complex lateral bending/torsion of the vertebral column.

E. Intervertebral disks. Comprise 20% of length of vertebral column. Each disk has outer annulus fibrosus, which is concentrically arranged wrapping of dense fibrous tissue. Deep to annulus fibrosus is softer nucleus pulposus, usually eccentrically placed so that the anterior edge of annulus fibrosus is thicker than posterior aspect of annulus fibrosus. Disks act as shock absorbers for vertebral column. They contain 70-80% water and thus allow disk to compress when bending. Disks tend to dehydrate when standing and rehydrate when lying. Rehydration becomes less efficient with aging and the nucleus pulposus tends to harden. Extrusion of the nucleus pulposus through the annulus fibrosus produces a herniated disk, which may press on nerve roots or spinal cord causing pain and loss of motor skills.

F. Ligaments of vertebral column.
1. Anterior longitudinal: attached to anterior aspect of vertebral bodies, extends from atlas to sacrum. Provides natural splinting action when fractures occur on anterior aspect of bodies of vertebrae.
2. Posterior longitudinal: attached to posterior aspect of vertebral bodies, thus within vertebral canal. Extends from atlas to sacrum. Holds vertebral column together during violent hyperflexion.
3. Ligamentum flava: joins contiguous borders of adjacent laminae.
4. Interspinous (deep) and supraspinous (superficial) connect adjacent spinous processes. Thickened supraspinal ligament in cervical region=ligamentum nuchae

G. Movement: Is a function of thickness of intervertebral disks and orientation of articular processes of vertebrae. In cervical region, much movement tolerated because of thick intervertebral disks and horizontal orientation of articular processes. At atlantoaxial (C1-C2) joint, chief movement is rotation ("no" movement of head); at atlanto-occipital joint, chief movement is flexion and extension ("yes" movement of head). Limited movement in thoracic region because of frontal orientation of articular processes, thin intervertebral disks, overlapping spinous processes and attachment of ribs and sternum. In lumbar region, thick intervertebral disks and sagittal orientation of articular processes allow wide range of flexion and extension.

H. Applied anatomy
1. Transverse ligament of Atlas: Rupture of this ligament may drive dens into spinal cord.
2. Fracture of the dens: may cause dislocation of the second cervical vertebra on the first cervical vertebra, which may produce transection of the spinal cord.
3. Obtaining CSF: The iliac crest peaks at the space between L3 and L4. The subarachnoid space can easily be reached at this level or L4-L5 by a needle inserted in the midline at these levels. There is no danger to the spinal cord in adults, since it terminates at L2.

4. Sacral hiatus: anesthetic may be injected through this opening around the roots of sacral and lower lumbar nerves without entering the subarachnoid space. This is epidural anesthesia.

II. Muscles of Back. Arranged into superficial, intermediate, and deep groups. Superficial and intermediate groups are innervated by ventral primary rami of spinal nerves even though they are dorsally located. This is because these muscles are derived from ventrolateral musculature that migrated secondarily to the back during development. The deep group of back muscles are native to the back, developmentally speaking, and are innervated by dorsal primary rami of spinal nerves.

A. Superficial group. Muscles in this group insert on the upper limb. Latissimus dorsi and trapezius form the first layer of muscles in this group; levator scapulae and rhomboid major and minor form a second layer.

B. Intermediate group. These muscles probably function in the mechanics of respiration. Two muscles in this group are the serratus posterior superior and serratus posterior inferior. These muscles are segmentally innervated by ventral primary rami of spinal nerves.

C. Deep group. These muscles are innervated by dorsal primary rami. Acting bilaterally, these muscles extend the vertebral column and regulate flexion of vertebral joints. Acting unilaterally, these muscles participate in lateral bending and rotation. Principal examples of this group are the splenius, erector spinae, semispinalis, multifidus, and rotatores.

D. Shoulder region muscles. These include the supraspinatus, infraspinatus, teres minor, subscapularis, teres major, and deltoid. These muscles all attach to the humerus and participate in various gleno-humeral movements.

E. Suboccipital muscles. These small muscles are associated with the suboccipital triangle and are four in number: inferior oblique, rectus major, superior oblique, and rectus minor. These muscles are innervated by the dorsal primary ramus of C_1 (suboccipital nerve).

III. Pectoral region.

A. Muscles and fascia. Muscles in this region include the pectoralis major and minor and the subclavius. The pectoralis major is a powerful adductor, medial rotator, and flexor of the arm. Pectoralis minor protracts the pectoral girdle. The subclavius cushions the clavicle and may help depress it. The subclavius and pectoralis minor are invested by the clavipectoral fascia. The double layered fascial continuation in the space between the two muscles is called the costocoracoid membrane. The costocoracoid membrane is perforated by the thoracoacromial artery, lateral pectoral nerve and cephalic vein.

B. Mammary gland. Located in superficial fascia and extending from the second to fourth intercostal spaces in the nulliparous female. The gland is described as a modified sweat gland. The glandular tissue (parenchyma) is composed of 15-20 lobules, each drained by a lactiferous duct that opens independently on the nipple. Near its termination, the lactiferous duct has a dilation, the lactiferous sinus. Well developed connective tissue trabeculae in the superficial fascia, termed Cooper's ligaments, support the parenchyma and extend superficially to attach to the skin of the breast. Arteries to the breast include medial mammary arteries, derived from the internal thoracic; lateral mammary arteries, derived from the lateral thoracic; pectoral branches from thoracoacromial artery and contributions from anterior intercostal arteries. The breast is segmentally innervated by intercostal nerves, since it is a gland that develops in the skin. Lymphatics from the breast drain primarily into axillary nodes. Skin lymphatics may also drain into the thorax (via lymphatics accompanying parasternal perforating vessels) or into the abdomen (via surface lymphatics caudally to the umbilicus, subsequently following the ligamentum teres into the abdomen) or connect to the opposite breast.

IV. Upper limb: Embryological considerations

   A. Limb rotation: After formation of limb buds, the axis of the limb rotates 90° laterally so that preaxial (flexor) musculature comes to lies anteriorly and postaxial (extensor) musculature ends up posteriorly.

   B. Innervation pattern: In the arm and forearm, preaxial muscles are innervated by preaxial nerves (musculocutaneous, median, ulnar); postaxial muscles are innervated by postaxial nerves (axillary, radial).

V. Axilla and contents: A pyramidal-shaped space bounded anteriorly by the muscular anterior axillary fold (pectoralis major and minor), posteriorly by the muscular posterior axillary fold (latissimus dorsi, teres major, subscapularis), medially by the thoracic wall including the overlying serratus anterior, and laterally by the intertubercular groove of the humerus. Contents of the axilla include fat and connective tissue, numerous lymph nodes, axillary artery and vein, and brachial plexus.

VI. Brachial plexus: The brachial plexus is an intermingling of several spinal cord ventral primary rami enabling the formation of peripheral nerves which contain several segments from the spinal cord. The brachial plexus is formed from C_5 to T_1 ventral primary rami. Often there are additional contributions from C_4 and T_2.

   A. Rami: the most proximal portion of the plexus, rami emerge between anterior and middle scalene muscles. The dorsal scapular nerve is a branch of C_5 and sometimes additional input from C_4 and innervates the two rhomboid muscles and the levator scapulae. The long thoracic nerve (C_{5-7}) innervates the serratus anterior.

   B. Trunks: the second stage in the formation of the plexus. The upper trunk is formed by joining of C_5 and C_6 rami. The suprascapular nerve and nerve to the subclavius arise from the upper trunk. The middle trunk is formed by lateral continuation of the ventral primary ramus of C_7. The inferior trunk is formed by joining of C_8 and T_1 rami.

   C. Divisions: Each trunk divides into anterior and posterior divisions. Anterior divisions of the upper and middle trunks unite to form the lateral cord ($C_{5,6,7}$). Anterior divisions of C_8 and T_1 unite to form the medial cord. Posterior divisions of all trunks unite to form the posterior cord.

   D. Cords: Cords are named according to their relationship to the axillary artery in the axilla. The lateral cord=lateral to artery, medial cord=medial to artery, and posterior cord=posterior to artery. Lateral cord gives off lateral pectoral nerve, musculocutaneous nerve, and a major contribution (lateral root) to the median nerve. Medial cord gives off the medial pectoral nerve, medial brachial cutaneous nerve, medial antebrachial cutaneous nerve, ulnar nerve, and the median root to the median nerve. The posterior cord gives off the upper subscapular nerve, the thoracodorsal nerve, the lower subscapular nerve, and the axillary and radial nerves.

E. "Big five" terminal branches: (1) musculocutaneous, from lateral cord, usually perforates the coracobrachialis muscle; (2) median, from medial and lateral cords; (3) ulnar, from medial cord; (4) radial, from posterior cord; and (5) axillary, from posterior cord.

VII. Dermatomes: Dermatomes of the upper extremity are important, particularly for diagnostic purposes. Review a diagram of dermatomes in one of the standard atlases.

VIII. Muscles of arm: Arm musculature is divided into flexor (preaxial) and extensor (postaxial) groups.

   A. Flexors: There are 3 flexors, all innervated by the musculocutaneous nerve: biceps brachii, brachialis, and coracobrachialis.

   B. Extensors: There are two extensors in the arm, both innervated by the radial nerve; the large triceps brachii and the small, often obscure, anconeus.

IX. Cubital fossa: An important triangular transition area on the anterior surface at the elbow. The base of this triangular area is an imaginary line connecting the two epicondyles of the humerus. The lateral boundary is the brachioradialis and the medial boundary is the pronator teres. The brachialis covers the floor of the fossa. In the middle of the fossa is the biceps tendon, which gives off the bicipital aponeurosis. Medial to the tendon is the brachial artery, and medial to the artery is the median nerve. The radial nerve lies deep to the brachioradialis in the cubital fossa. Superficial to the bicipital aponeurosis one can usually find the median cubital vein, important clinically for intravenous injections and drawing blood.

X. Muscles of forearm: Forearm muscle nomenclature may appear complex, but the names of the muscles usually reflect the insertion or function of the muscles.

   A. Superficial flexors and pronators. These include the pronator teres, flexor carpi radialis, palmaris longus (flexes at wrist) and flexor carpi ulnaris. All of these muscles are innervated by the median nerve, except flexor carpi ulnaris which is innervated by the ulnar nerve.

   B. Deep flexors and pronators. These include flexor digitorum superficialis, flexor pollicis longus, flexor digitorum profundus and pronator quadratus. These muscles all receive innervation from the median nerve. The flexor digitorum profundus is a dually innervated muscle: the part of the muscle attaching to the index and middle finger is innervated by the median nerve. The part of the muscle attaching to the ring and little finger is innervated by the ulnar nerve.

   C. Extensors and supinators. These include brachioradialis, extensor carpi radialis longus, extensor carpi radialis brevis, supinator, extensor digitorum, extensor digiti minimi, extensor carpi ulnaris, abductor pollicis longus, extensor pollicis brevis, extensor pollicis longus and extensor indicis. All muscles in the extensor-supinator group are innervated by the radial nerve.

XI. Hand. For purposes of description, the hand is divided into compartments, separated by fascia. Compartments include the thenar or thumb compartment, the hypothenar, or little finger compartment, the central compartment and the deep interosseous-adductor compartment.

   A. Thenar compartment: contains three intrinsic muscles: abductor pollicis brevis, flexor pollicis brevis and opponens pollicis. These muscles are all innervated by the recurrent branch of the median nerve. In addition, the deep portion of flexor pollicis brevis (portion deep to tendon of flexor pollicis longus tendon) may receive innervation from the ulnar nerve. The major arterial supply is the princeps pollicis, a branch of the deep palmar arterial arch. The superficial branch of the radial artery, which completes the superficial palmar arterial arch, also contributes.

B. Hypothenar compartment: contains three intrinsic muscles: abductor digiti minimi (quinti), flexor digiti minimi, and opponens digiti minimi. These muscles are all innervated by the ulnar nerve. The arterial supply is from the ulnar artery or superficial palmar arterial arch.

C. Central compartment: Underlies palmar aponeurosis and contains tendons of flexor digitorum superficialis and profundus and lumbrical muscles. Lumbricals are small worm-like muscles (four) that arise from the radial side of the tendon of flexor digitorum profundus and insert into the radial border of the extensor expansion. Because of their insertion into the extensor expansion, they flex at the MP joints and extend the interphalangeal joints. The two radial-most lumbricals (#1 and #2) are innervated by the median nerve, lumbricals 3 and 4 are innervated by the ulnar nerve. Recall that flexor digitorum superficialis is innervated by the median nerve, but flexor digitorum profundus is innervated by both the median and ulnar nerve. The part of flexor digitorum profundus that supplies the tendons going to the ring and little finger is innervated by ulnar nerve, the part of the muscle that supplies tendons going to the index and middle finger is innervated by the median nerve. The central compartment is vascularized by branches of the superficial palmar arterial arch.

D. Interosseous-adductor compartment: The deep part of the hand that contains adductor pollicis, 3 palmar interosseous muscles and 4 dorsal interosseous muscles. The fourth palmar interosseous muscle (for index finger) is incorporated into the adductor pollicis. The reference finger for abduction and adduction is the middle finger. Palmar interossei are adductors, and dorsal interossei are abductors with reference to this finger. All muscles in this compartment are innervated by the ulnar nerve (deep branch).

E. Summary of innervation: All intrinsic muscles of the hand are innervated by the ulnar nerve, except the three muscles of the thenar eminence and the first and second lumbricals, which receive the median nerve. Sensory on palmar side is provided by median and ulnar nerves. The median is sensory on the radial side of the thumb and thenar eminence, laterally across the palm to the 4th metacarpal, and over the index, middle and radial half of ring fingers. The ulnar is sensory on the ulnar half of ring finger and little finger, skin over hypothenar eminence, and it continues dorsally to serve all of the dorsum of the little finger and the ulnar half of the ring finger. The ulnar is also sensory to the dorsum of the hand overlying the fourth and fifth metacarpals. The remainder of the dorsum of the hand and thumb, index, middle, and radial half of ring finger are provided sensation by the radial nerve.

XII. Functional anatomy of upper limb.

A. Movements at shoulder: includes movement of the scapula and movements at the glenohumeral joint. Movements of the scapula include elevation, depression, upward rotation, downward rotation, and protraction and retraction. Movements at the glenohumeral joint include flexion of the arm, extension of the arm, abduction of the arm, adduction of the arm, and medial and lateral rotation.

B. Movements at elbow: flexion and extension (humeroulnar) and pronation and supination, accomplished by the radius rotating around the ulna.

C. Movements at wrist: These include flexion, extension, abduction (radial deviation) and adduction (ulnar deviation).

D. Movements of the hand: Obviously the hand pronates and supinates when the radius rotates around the ulna. Abduction and adduction of the fingers has already been mentioned. After the long flexors of the digits have flexed at the proximal or distal interphalangeal joint, continued contraction will produce flexion at the MP joints and wrist. Flexion at the MP joint is also a function of the lumbricals. Extension at the interphalangeal joints is accomplished by the extensor digitorum and lumbricals. Lumbricals extend the interphalangeal joints by virtue of the fact that they insert into the extensor expansion, which is on the dorsal side of these joints. Thumb movements are unique. Extension=movement of the thumb away from the hand in the plane of the hand. Flexion is the opposite movement. Abduction=movement of the thumb up and out of the plane of the hand and adduction brings the thumb back toward the plane of the hand. Opposition=drawing the thumb near the center of the palm and opposing the tips of other fingers.

E. Nerve lesions.
   1. Ulnar: Lesion at wrist, distal to where nerve innervates the ulnar half of flexor digitorum profundus, produces the classic ulnar claw hand. The claw results from paralysis of lumbricals 3 and 4 while the innervation to the profundus is still intact. Since the interossei are also affected, abduction and adduction of the fingers is eliminated. If the ulnar nerve has a lesion proximal to where it innervates the flexor carpi ulnaris and ulnar half of flexor digitorum profundus, there is muscle wasting on the medial forearm and hypothenar eminence, but the claw hand is not well developed because the ulnar half of the flexor digitorum profundus in addition to lumbricals 3 & 4 are flaccid and unable to contract. Abduction and adduction of the fingers is eliminated.
   2. Median: Lesion of the median recurrent nerve produces thenar wasting and an inability to have smooth opposition. There is also sensory loss.
   3. Radial: Lesion of this nerve around the surgical neck of the humerus produces wrist drop.
   4. Upper trunk brachial lesion: (Erb-Duchenne paralysis) produces "waiter's tip" syndrome. Upper limb is medially rotated and extended because lateral rotators and flexors are compromised.

XIII. Vasculature of upper limb.

A. Subclavian artery: becomes the axillary artery after crossing the first rib. Divided into three parts by the anterior scalene muscle which crosses anterior to the artery. Three parts of the artery are defined by their relationship to the anterior scalene: a part medial (first part), a part deep to the muscle (second part), and a part lateral to the muscle (third part).
   1. First part: gives off vertebral, thyrocervical, and internal thoracic arteries.
   2. Second part: gives off costocervical trunk.
   3. Third part: gives off dorsal scapular artery.

B. Axillary artery: The continuation of the subclavian artery. Begins at the lateral border of the first rib and continues to the lower border of teres major. The tendon of pectoralis minor crosses it anteriorly and divides the vessel into 3 parts: a part medial to the pectoralis minor tendon (first part); a part behind the tendon of pectoralis minor (second part) and a part lateral to the tendon of pectoralis minor (third part).
   1. First part: gives off one branch, the superior thoracic artery.
   2. Second part: gives off two branches, the thoracoacromial and lateral thoracic arteries.
   3. Third part: gives off three branches, the anterior humeral circumflex, posterior humeral circumflex and subscapular arteries.

C. Brachial artery: The continuation of the axillary artery. At the lower border of teres major, the axillary artery becomes the brachial artery. The brachial artery continues into the cubital fossa where it divides into the ulnar and radial arteries. Branches of the brachial artery include the profunda brachii (deep brachial), superior ulnar collateral, and inferior ulnar collateral.

D. Arterial supply to forearm and hand.
1. Radial artery: gives off radial recurrent which ascends around the lateral aspect of the elbow and helps form the elbow arterial anastomosis. Also, usually a superficial palmar branch is given off distally before the radial artery goes into the anatomical snuff box. This branch anastomoses with and completes the superficial palmar arterial arch. The main trunk of the radial artery continues into the floor of the anatomical snuff box and enters the hand by passing between the two heads of the first dorsal interosseous muscle and continues medially between the transverse and oblique heads of adductor pollicis, to join the deep branch of the ulnar artery and complete the formation of the deep palmar arterial arch. Thus the radial artery is the primary contributor to the deep palmar arterial arch. The princeps pollicis and radialis indicis are major branches from the deep palmar arterial arch.
2. Ulnar artery: Usually the larger branch of the brachial artery which continues down the ulnar side of the forearm. Branches include the anterior and posterior ulnar recurrent arteries and the common interosseous artery which gives off an anterior and posterior branch. The ulnar artery continues into the hand, and forms the superficial palmar arterial arch. Digital arteries arise from this arch.

E. Collateral circulation around joints.
1. Shoulder: Branches of the thyrocervical trunk (from subclavian artery) and subscapular (from third part of axillary) arteries, along with intercostal arteries, form anastomoses around the shoulder which may serve as a collateral route for blood to flow to the upper extremity.
2. Elbow: Recurrent branches from the ulnar anastomose with superior and inferior ulnar collateral branches from the brachial artery and branches from the deep brachial anastomose with the radial recurrent artery. This anastomoses forms an effective collateral route around the elbow.

F. Veins of the upper limb.
1. Superficial: The cephalic system originates on the radial side of the hand and the basilic system originates on the ulnar side of the hand. There are numerous connections between these two major vessels in the forearm and a specific pattern is not apparent. In the arm, the basilic vein is on the medial side and the cephalic vein is on the lateral side. The cephalic continues superiorly to pierce the costocoracoid membrane and drain into the axillary. In the proximal arm, the basilic converges with the brachial veins to form the axillary vein.
2. Deep. Deep veins, usually 2, accompany deep muscular arteries in the forearm and arm. They are often referred to as vena comitantes of the named artery.

XIV. Wrist.

A. Osteology: There are eight carpal bones arranged in a proximal and distal row. The bones of the proximal row, beginning on the radial side are: scaphoid, lunate, triquetrum and pisiform. Bones of the distal row, beginning on the radial side are: trapezium, trapezoid, capitate, and hamate. The scaphoid is the most frequently fractured carpal bone and the lunate is the most frequently dislocated. The capitate is a major carpal bone which transmits most of the force from the hand superiorly into the forearm. The carpal bones are held together by numerous dorsal and palmar ligaments.

B. Tendons and nerves: The flexor retinaculum is attached to the palmar surface of the carpal bones and forms a fibro-osseous (carpal) tunnel through which tendons of the flexor digitorum muscles, and flexor pollicis longus pass. The median nerve also passes through this tunnel. Pressure on the median nerve as it passes through this tight space causes carpal tunnel syndrome.

XV. Joints.

A. "Scapulo-thoracic joint." This is not a joint in the conventional sense but instead refers to the rotary movement of the scapula on the dorsal scapula wall. For instance, in complete abduction of the upper extremity, the scapula rotates at least 60^0 upward, in coordination with movement at the glenohumeral joint, to position the glenoid surface of the scapula.

B. Sternoclavicular. The point of bony union between the trunk and upper limb. The synovial cavity has a fibrocartilaginous disk interposed in the joint space. The joint is reinforced anteriorly and posteriorly by sternoclavicular ligaments.

C. Acromioclavicular. This synovial joint connects the lateral end of the clavicle with the acromion of the scapula. Though not a part of the joint, the coracoclavicular ligaments (conoid and trapezoid) lend stability to the joint by holding the lateral end of the clavicle down, preventing the acromion from being forced deep to the opposing lateral end of the clavicle.

D. Shoulder (glenohumeral). A ball and socket with a large ball and very small bony socket and loose articular capsule. This allows for free mobility but very little stability. During abduction, the head of the humerus is literally held in place against the glenoid surface of the scapula by the short tendons of the rotator cuff muscles (supraspinatus, infraspinatus, teres minor, and subscapularis). The bony socket of the joint is deepened by a fibrocartilaginous rim, the glenoid labrum.

E. Elbow. A hinge joint between the humerus and ulnar. The proximal radioulnar articulation also occurs here. The joint is reinforced by strong collateral ligaments on each side (ulnar and radial collateral). The radial head is held against the radial notch of the ulna by the strong annular ligament of the radius.

F. Wrist (radiocarpal articulation). The dominant long bone at the distal radiocarpal articulation is the enlarged radius. The scaphoid and lunate are dominant carpal bones at this articulation. The wrist allows abduction, adduction, flexion, extension and circumduction.

XVI. Radiology of upper limb.

A. Tear of rotator cuff tendons: causes an abnormal communication between the joint space and the subacromial bursa. Injection of contrast agent into the joint space and subsequent examination will confirm these tears.

B. Shoulder dislocations: most are inferior and/or anterior.

C. Acromioclavicular (AC) joint: With trauma, this joint may separate. When this occurs, usually the strong conoid and trapezoid ligaments are torn.

D. Elbow joint: Usually the faint shadows of the fat pads in the coronoid and radial fossae can be seen. Trauma to the joint such as radial head fracture, usually displaces these fats pads anteriorly.

E. Wrist: The scaphoid is the most frequently fractured bone in the wrist. Often the blood supply is not adequate in both ends of the bone, and the end of the broken bone without adequate blood supply may undergo aseptic necrosis.

HEAD AND NECK

I. Overview of embryology. The organizational plan of head and neck anatomy is greatly aided by an understanding of basic aspects of head and neck embryology, particularly the branchial apparatus. Six branchial arches, numbered cranially to caudally, first become recognizable during the fourth week as neural crest cells migrate into the head and neck region. The fifth arch is rudimentary or absent in the human. Associated with each branchial arch is an artery, a nerve, a cartilaginous component, and a muscular component. These components contribute extensively to the definitive anatomy of the face, neck, nasal cavities, mouth, larynx and pharynx.

A. Derivatives of branchial arch cartilaginous components. First arch: malleus and incus (of middle ear), anterior ligament of malleus, and sphenomandibular ligament. Second arch: stapes (of middle ear), styloid process of temporal bone, stylohyoid ligament, lesser cornu and superior part of body of the hyoid bone. Third arch: greater cornu and inferior part of body of hyoid bone. Fifth arch: rudimentary or absent. Fourth and Sixth arches: all laryngeal cartilages except the epiglottic cartilage.

B. Muscles derived from branchial arches: First arch: muscles of mastication (masseter, lateral and medial pterygoids, temporalis), anterior belly of digastric, mylohyoid, tensor tympani, and tensor veli palatini. Second arch: muscles of facial expression, posterior belly of digastric, stapedius, and stylohyoid. Third arch: one muscle, the stylopharyngeus. Fourth arch: circular constrictors and longitudinal muscles of the pharynx, levator veli palatini, muscularis uvulae, and cricothyroid. Sixth arch: all intrinsic muscles of the larynx (except cricothyroid) and cricopharyngeus. The sixth arch may also contribute to the trapezius and sternocleidomastoid, however, these muscles receive their motor innervation from the spinal root of the accessory nerve.

C. Nerves associated with branchial arches: First arch: trigeminal nerve. Second arch: facial nerve. Third arch: glossopharyngeal nerve. Fourth arch: superior laryngeal branch of vagus. Sixth arch: recurrent laryngeal branch of vagus.

II. Cervical plexus. The cervical plexus is formed from ventral primary rami of spinal nerves C_{1-4}. The plexus is deep to the sternocleidomastoid muscle and provides motor supply to neck muscles derived from somites and the diaphragm, and sensation to the neck.

A. Motor branches.
    1. C_1 ventral primary ramus has a descending branch, the superior ramus of ansa cervicalis, which travels inferiorly and medially with the hypoglossal nerve. Medially, a branch of the superior ramus (C_1) innervates the geniohyoid and thyrohyoid muscles.
    2. The superior ramus of ansa cervicalis (C_1), joins the inferior ramus ($C_{2,3}$) to form a nerve loop (ansa cervicalis) around (usually) the internal jugular vein. The ansa cervicalis complex provides motor innervation to the remaining 3 infrahyoid muscles (sternohyoid, omohyoid, sternothyroid).

67

3. Phrenic nerve. This nerve innervates the diaphragm. It is primarily C_4, although it usually receives additional contributions from C_3 and C_5.
4. Direct muscular branches from C_{1-4} supply prevertebral muscles (longus capitis, longus coli, etc).

B. Sensory branches.
1. C_1 gives a sensory branch which ascends superiorly through the foramen magnum to innervate the meninges.
2. C_2 and C_3 converge to form the following sensory nerves: lesser occipital, greater auricular, and transverse cervical.
3. C_3 and C_4 converge to form the supraclavicular nerves.
4. C_{2-4} provide sensory branches which join the accessory nerve to provide sensory supply to the sternocleidomastoid ($C_{2,3}$) and trapezius ($C_{3,4}$).

III. Cranial nerves. Twelve cranial nerves, part of the peripheral nervous system, exit from the cranium. These nerves are not uniform in their composition and each must be considered as a "special case." In addition to the four general functional components (GSE, GSA, GVE, GVA) found in all spinal nerves, cranial nerves may have an additional three "special" functional components. Specialized senses account for two special afferent components: special visceral afferent (SVA, taste and smell); special somatic afferent (SSA, sight, equilibrium and hearing). Some authors do not distinguish between SVA and SSA, instead referring to all these modalities as special sensory (SS). Skeletal muscles which develop in association with branchial arches (branchiomeric muscles) are classically considered to be of visceral origin; fibers which innervate these muscles are called special visceral efferent (SVE). Recent evidence suggests that these muscles develop from paroxial mesoderm like all other skeletal muscle (see musculoskeletal system in Embryology section).

A. Olfactory nerve (cranial nerve I). The nerve of smell. Exits the cranium through the cribriform plate of the ethmoid bone. One functional component: SVA, smell.

B. Optic nerve (cranial nerve II). The nerve of sight. Exits the cranium through the optic canal (sphenoid bone). One functional component: SSA, sight.

C. Oculomotor nerve (cranial nerve III). Innervates 4 extraocular muscles and the levator palpebrae superiorus, and provides parasympathetic autonomic innervation to the orbit (sphincter pupillae and ciliary body). Two functional components: GSE and GVE. The nerve emerges from the ventral brain and passes between the superior cerebellar and posterior cerebral arteries, continues lateral to the posterior communicating artery and descends into the lateral wall of the cavernous sinus, superior to the trochlear nerve. In the cavernous sinus, GSA components from the ophthalmic division of the trigeminal nerve and GVE sympathetic components from the internal carotid plexus "ride" with III for peripheral distribution. The III nerve enters the orbit through the superior orbital fissure, and divides into a superior and inferior division.
1. Superior division: GSE components innervate 2 muscles derived from myotomes: the levator palpebrae superiorus and the superior rectus.
2. Inferior division: GSE components innervate 3 muscles derived from myotomes: medial rectus, inferior rectus, and inferior oblique. The mnemonic for muscles innervated by the III nerve is 35X. The parasympathetic GVE component of III travels with the inferior division of III to reach the ciliary ganglion, suspended from the nasociliary nerve near the apex of the orbit, for synapse. The ciliary ganglion contains the perikarya of postganglionic parasympathetic GVE components. Sympathetic GVE components, and

sensory components, pass through the ganglion. Postganglionic parasympathetic GVE components as well as sympathetic GVE and sensory (GSA) components leave the ciliary ganglion via the short ciliary nerves for peripheral distribution. Parasympathetic postganglionic GVE components innervate smooth muscle of the ciliary body and the pupillary sphincter. Sympathetic GVE components innervate smooth muscle which dilates the pupil. Sensory (GSA) components of the short ciliary nerves provide sensation to the cornea.

D. Trochlear nerve: (cranial nerve IV). One functional component: GSE. Innervates one extraocular muscle, the superior oblique. The nerve leaves the dorsum of the brain stem, decussates, and emerges on the ventral surface of the brain posterior to the posterior clinoid process. It then passes forward in the lateral wall of the cavernous sinus inferior to the III nerve. The nerve reaches the orbit via the superior orbital fissure (sphenoid bone) and innervates one muscle derived from myotomes: the superior oblique. The mnemonic for the trochlear nerve is SO4.

E. Trigeminal nerve: (cranial nerve V). Two functional components: GSA and SVE. The great sensory nerve of the anterior face and the nerve that innervates branchiomeric muscles from the first arch.
   1. Trigeminal ganglion. The large sensory ganglion of the V nerve that lies deep to dura, near the apex of the petrous portion of the temporal bone. Three major branches (divisions) emerge from this ganglion: the ophthalmic division (V^1), maxillary division (V^2), and mandibular division (V^3).
   2. Ophthalmic division. Purely sensory (GSA). Leaves the trigeminal ganglion, travels anteriorly in the wall of the cavernous sinus, and reaches the orbit via the superior orbital fissure. In the orbit, it divides into 3 nerves: the frontal, lacrimal and nasociliary.
   3. Maxillary division. Purely sensory (GSA). Like the ophthalmic branch, it leaves the trigeminal ganglion, travels anteriorly in the wall of the cavernous sinus, leaves the cranium through the foramen rotundum (sphenoid), traverses the pterygopalatine fossa, and passes through the inferior orbital fissure and becomes the infraorbital nerve. It provides GSA components to the sphenoid, ethmoid, and maxillary paranasal sinuses, maxillary teeth and gums, part of the nasopharynx and hard palate, and skin of the anterior zygomatic and anterior temporal region.
   4. Mandibular division. A mixed division (GSA and SVE). It leaves the cranium via the foramen ovale (sphenoid), gives off a branch which reenters the cranium to supply the dura. The mandibular nerve supplies sensation to the teeth and gums of the lower jaw, anterior two-thirds of the tongue, buccal mucosa, and it has a large cutaneous distribution around the mandible and via the auriculotemporal nerve to the external auditory meatus, parotid gland, and temporomandibular articulation. SVE components supply muscles of mastication and several other muscles derived from first arch musculature.

F. Abducens nerve (cranial nerve VI). One functional component, GSE, and the nerve innervates one extraocular muscle, the lateral rectus. The nerve leaves the brain and continues anteriorly through the cavernous sinus and reaches the orbit via the superior orbital fissure (sphenoid).

G. Facial nerve (cranial nerve VII). The great motor nerve to the face. Five functional components: SVE to muscles of facial expression and other muscles derived from the second arch; GVE=parasympathetics to lacrimal (via synapse in pterygopalatine ganglion) and submandibular gland and sublingual gland (via synapse in submandibular ganglion);

SVA=taste from anterior two-thirds of tongue, GVA= sensory from a small lateral area of soft palate, and GSA=sensory from skin on posterior surface of external ear. The geniculate ganglion lies in the genu of the facial canal and contains nerve cell bodies of all three sensory components. The facial nerve enters the internal auditory meatus as two nerves: the large branchiomeric portion (the SVE motor root) and the smaller intermediate nerve, which contains the sensory and autonomic components.

H. Vestibulocochlear (cranial nerve VIII). One functional component, SSA, which mediates equilibrium and hearing. The nerve exits the cranium via the internal auditory meatus, along with the VII nerve.

I. Glossopharyngeal (cranial nerve IX). Exits the skull via the jugular foramen. Five functional components: GVA=general sensory from posterior one third of tongue mucosa, lateral pharynx in region of palatine tonsils, carotid body and sinus, and mucous membrane of tympanic cavity; SVA=taste from posterior one third of tongue; SVE=motor to stylopharyngeus muscle, GSA=sensory from external auditory meatus and small portion of external ear; GVE=parasympathetic autonomic to parotid gland via synapse in the otic ganglion. There are two sensory ganglia of the IX nerve: an inferior (petrosal) ganglion contains GVA and SVA nerve cell bodies, and the superior ganglion contains GSA nerve cell bodies.

J. Vagus nerve (cranial nerve X). Exits the skull via the jugular foramen. Five functional components: GVE=parasympathetic to thoracic and abdominal viscera; SVE= branchiomotor to muscles of pharynx derived from the fourth and sixth arches; GVA= sensory from a small area of the pharynx, from the carotid body, from the laryngeal mucosa and from viscera of body cavities superior to the pelvis; SVA=taste from a few taste buds on the epiglottis; GSA=sensory from external auditory meatus, external aspect of tympanic membrane, and part of the pinna of the ear. There are two sensory ganglia of the X nerve: an inferior (nodose) ganglion contains nerve cell bodies of GVA and SVA components and the superior (jugular) ganglion contains nerve cell bodies of GSA components.

K. Accessory nerve (cranial nerve XI, spinal root). Exits the cranium through the jugular foramen. One functional component: SVE, to sternocleidomastoid and trapezius. The spinal root is derived from a convergence of contributions from the upper five cervical segments of the spinal cord. Some descriptions include a bulbar root of XI which contains SVE components. However, these bulbar contributions distribute peripherally with branches of the X nerve and are considered to be a part of the X nerve in the current description.

L. Hypoglossal (cranial nerve XII). One functional component, GSE. The hypoglossal provides voluntary motor innervation to the three extrinsic muscles (hyoglossus, genioglossus, styloglossus) of the tongue. The nerve leaves the cranium via the hypoglossal canal (occipital bone), proceeds anteriorly and inferiorly and ascends deep to the posterior belly of the digastric muscle to reach the tongue.

IV. Superficial face. Important structures include the muscles of facial expression, derived from the second arch, numerous branches of the facial nerve innervating these muscles, the facial artery and vein, and the parotid gland and duct.

A. Facial artery. A branch of the external carotid. Ascends across the mandible anterior to the masseter muscle onto the face. On the face, it is distinctively tortuous. Gives off superior and inferior labial branches, continues superiorly as the lateral nasal and terminates as the angular artery in the medial canthus of the eye.

B. Facial vein. Accompanies the facial artery and drains into the jugular system. This valveless vein is connected indirectly to the important cavernous sinus via the ophthalmic vein. There are also communications with the deep pterygoid plexus of veins. These are anatomical routes by which infections may spread from the superficial to deep face or into the dural venous sinuses.

C. Facial nerve. Branchiomeric components exit the stylomastoid foramen (temporal bone) and enter the parotid gland. Within the gland, the nerve divides into two major divisions and each division gives off several branches. These branches (temporal, zygomatic, buccal, mandibular, cervical) emerge around the superior and anterior edge of the parotid gland and innervate muscles of facial expression.

D. Parotid gland. The largest salivary gland. Sits in front of the ear, and covered by a dense fascia. Receives its parasympathetic secretomotor innervation from the IX nerve via postganglionics from the otic ganglion which accompany the auriculotemporal branch of the mandibular nerve. The auriculotemporal nerve also conveys sympathetic fibers, derived from the plexus around the middle meningeal artery, to the gland.

V. Temporal region. The superficial temporal artery, branch of the external carotid, ascends anterior to the ear, accompanied by the temporal branch of the auriculotemporal nerve (from mandibular nerve). The temporalis muscle descends deep to the zygomatic arch to insert on the coronoid process of the mandible. The masseter muscle, originates from the zygomatic arch and descends to insert on the ramus of the mandible. Both these powerful muscles of mastication are innervated by the trigeminal nerve (mandibular branch).

VI. Infratemporal region. This part of the deep face contains two important muscles of mastication, and major nerves and vessels. These structures are contained in the infratemporal fossa, a space bounded laterally by the ramus of the mandible and medially by the lateral pterygoid plate (sphenoid).

A. Temporomandibular joint. This joint borders the infratemporal region, posteriorly. This is a synovial joint containing an articular disk which divides the joint cavity into two spaces: The upper space allows a gliding motion between the disk and mandibular fossa, such as when protruding the jaw. The lower space allows hinge movement between the mandibular condyle and disk, such as when opening the mouth. This complex joint receives sensory innervation from the auriculotemporal nerve.

B. Muscles. Two important muscles in this region participate in mastication and are, of course, innervated the mandibular branch of the V cranial nerve. The lateral pterygoid takes origin primarily from the lateral pterygoid plate and inserts into the mandibular neck and articular disk of the temporomandibular joint. This muscle is important for protrusion of the mandible. The medial pterygoid muscle runs a course on the medial side of the mandible similar to the masseter muscle on the lateral side. Thus it is important in grinding and chewing. Its anterior fibers may also participate in protrusion of the jaw.

C. Mandibular division of trigeminal nerve. This division is a mixed nerve, while the ophthalmic and maxillary divisions are purely sensory. The mandibular nerve enters the infratemporal fossa via the foramen ovale and gives motor branches to the four muscles of mastication. It also provides the following sensory branches: meningeal, sensory to dura mater; buccal, sensory to buccal mucosa; auriculotemporal, sensory to external ear and scalp; lingual, sensory to anterior two-thirds of tongue; and inferior alveolar, sensory to lower jaw.

D.   Chorda tympani. This is a branch of the VII nerve which mediates taste from the anterior two-thirds of the tongue and carries parasympathetic preganglionic fibers to the submandibular ganglion. High in the infratemporal fossa, this nerve exits the petrotympanic fissure and joins the lingual branch of the mandibular nerve for peripheral distribution.

E.   Maxillary artery. This branch of the external carotid enters the infratemporal fossa by passing posterior to the neck of the mandible. Important branches of the maxillary artery include numerous branches which supply muscles of mastication; the middle meningeal artery, which supplies the major portion of the periosteal dura in the cranium; inferior alveolar, which supplies the lower jaw; the posterior superior alveolar, which supplies the upper jar; the descending palatine, which supplies the hard palate; and the sphenopalatine, which supplies the nasal cavity.

VII.   Floor of mouth. The tongue is attached to the floor of the mouth and several important nerves, muscles, and vessels are in this region.

A.   Muscles. From superior to inferior, these include the genioglossus, which protrudes the tongue; geniohyoid, which stabilizes the hyoid; mylohyoid, which stabilizes the hyoid and depresses the mandible; and anterior belly of digastric, which also stabilizes the hyoid.

B.   Tongue. This structure has three extrinsic muscles and a complex innervation. The styloglossus, retracts the tongue; hyoglossus, flattens the tongue from side to side; and genioglossus, protrudes the tongue. Five nerves innervate the tongue: trigeminal, general sensation from the anterior two thirds; facial, taste from anterior two thirds; glossopharyngeal, taste and general sensation from posterior one third; vagus, motor to palatoglossus, which elevates back of tongue; and hypoglossal, motor to genioglossus, hyoglossus and styloglossus.

C.   Lingual artery. Branch of external carotid which lies deep to the hyoglossus muscle and ascends the posterior border of this muscle to enter the floor of mouth. Often arises in common with the facial artery.

D.   Nerves. The hypoglossal and lingual nerves lie deep to the tongue. Details of these nerves have been previously mentioned.

E.   Sublingual and submandibular salivary glands. The sublingual gland, the deep part of the submandibular gland, and the submandibular duct are deep to the tongue in the floor of the mouth.

VIII.   Paranasal sinuses. There are four groups of these sinuses; all are lined with a richly vascularized mucosa. These sinuses also receive an abundant sensory innervation from the trigeminal nerve (V^1 and V^2). All paranasal sinuses drain into the nasal cavity or nasopharynx. The four groups of sinuses are: frontal, maxillary, sphenoid, and ethmoid air cells.

IX.   Cranial dura mater and venous sinuses. In the cranium, the tough dura consists of two layers: an outer layer which serves as periosteum for bone (periosteal layer) and an inner layer which is applied to the brain (meningeal layer). The meningeal layer forms structures which help support and separate various lobes and hemispheres of the brain. Between duplications of the meningeal layer or between meningeal and periosteal dura are dural venous sinuses. These sinuses are lined with endothelium and directly or indirectly drain into the internal jugular vein.

A.   Dural derivatives which support/separate portions of brain:

1. Falx cerebri: Sickle-shaped double layer of meningeal dura that projects inferiorly and separates the two cerebral hemispheres. Its lower free border contains the inferior sagittal sinus and the superior sagittal sinus is at the superior border.
2. Tentorium cerebelli: A double layer of meningeal dura, shaped like a tent, which serves to support the occipital lobes of cerebral hemispheres and separate these structures from the cerebellar hemispheres. The medial portion of the tentorium is free around the brain stem. The tentorium is attached posterolaterally to the occipital bone and anteriorly to the petrous process of the temporal bone.
3. Falx cerebelli: A poorly developed double layer of meningeal dura which separates the cerebellar hemispheres.
4. Diaphragma sellae: A double layer of meningeal dura that covers the hypophyseal fossa superiorly, and thus the hypophysis, except for an opening that allows the pituitary stalk to pass through.

B. Dural venous sinuses: These include the large superior sagittal sinus between the meningeal and periosteal dura along the midline; this is the site of CSF resorption into venous blood via the arachnoid granulations; the inferior sagittal sinus, at the free border of the superior sagittal sinus; the straight sinus, the continuation of the inferior sagittal sinus which receives the great cerebral vein; the confluens of sinuses, which receives the straight, superior sagittal and occipital sinuses; the transverse sinus, the lateral continuation from the confluens; and the sigmoid sinus, which is the continuation of the transverse sinus into the internal jugular bulb. Smaller sinuses include the sphenoparietal sinus, on the lesser wing of the sphenoid; the superior and inferior petrosal sinuses on the same aspects of the petrous process and the very important cavernous sinus with several nerves and the internal carotid artery passing through. Emissary veins pass through the skull and connect with external veins, and thus provide an anatomical route by which infection may pass from the surface deep into the brain.

X. Organization of neck: fascial planes.

A. Visceral. This surrounds the two visceral tubes in the neck. The anterior portion of this fascia is called pretracheal, the posterior portion around the pharynx is bucopharyngeal.

B. Prevertebral. This surrounds the vertebrae and prevertebral muscles. The potential space between the visceral and prevertebral fascia is the retropharyngeal space.

C. Carotid sheath. The fascial sheath that invests the carotid artery, internal jugular vein, and vagus nerve.

D. Investing fascia. As the name suggests, this fascia surrounds all neck structures.

XI. Triangles of neck: The neck is a complex area and for descriptive purposes, the neck is divided into several triangles. The sternocleidomastoid muscle divides the neck into posterior and anterior triangles, which are posterior and anterior to the reference muscle, respectively.

A. Posterior triangle: area between the sternocleidomastoid and trapezius muscles. Important structures found in this area include accessory nerve, external jugular vein, cutaneous branches of cervical plexus, transverse cervical and suprascapular arteries (from thyrocervical trunk), brachial plexus, and the anterior scalene muscle with overlying phrenic nerve.

B. Anterior triangle: area anterior to sternocleidomastoid over to midline of the neck. This triangle is divided into several triangles by smaller muscles crossing and bordering the larger anterior triangles. Smaller triangles include the muscular triangle of the neck, submental triangle (under the chin), submandibular triangle and carotid triangle.

XII. Vasculature of head and neck.

A. Branches of arch of aorta: From right to left, these include: brachiocephalic, left common carotid, and left subclavian. On the right side, the brachiocephalic subsequently divides into the right subclavian and right common carotid arteries.

B. Subclavian artery: Crosses posterior to anterior scalene muscle. Medial to the anterior scalene muscle it has 3 branches: vertebral, thyrocervical and internal thoracic. Posterior to the anterior scalene, it gives off the costocervical trunk. Lateral to the anterior scalene, it gives off the dorsal scapular artery.

C. Common carotid artery. Ascends in the neck, and in the carotid triangle divides into external and internal branches. The internal carotid artery is posterior to the external carotid artery and has no branches outside the cranium. In contrast, the external carotid artery has 8 branches.
   1. External carotid: The first branch is the superior thyroid, and the remaining branches in order are: lingual, facial, ascending pharyngeal, occipital, posterior auricular, and the terminal bifurcation forms the superficial temporal and maxillary arteries.
   2. Internal carotid: enters the cranium via the carotid canal (temporal bone), traverses the cavernous sinus, and immediately gives off the ophthalmic artery. Subsequently, it bifurcates into its two terminal branches, the anterior and middle cerebral arteries.

D. Circulus arteriosus (Circle of Willis). This arterial circle around the base of the brain is formed from contributions from the internal carotids anteriorly, and the vertebral arteries posteriorly. The vertebrals, branches of the subclavian, ascend through the foramen magnum and fuse on the anterior surface of the brainstem, forming the basilar. Branches from the vertebrals and basilar serve the cerebellum and pons. The basilar terminates by dividing into two posterior cerebral arteries. Posterior communicating arteries connect the posterior cerebrals to the internal carotid. Anterior carotids are connected by an anterior communicating artery. The arterial circle so formed helps to equalize blood flow to various parts of the brain.

E. Internal jugular vein. The continuation of the sigmoid dural venous sinus extra-cranially via the jugular foramen. A dilation, the superior jugular bulb, occurs at its origin. This major venous trunk has several communications with the external jugular system and ends by joining the subclavian vein to form brachiocephalic vein on each side.

F. External jugular vein. Lies deep to platysma but superficial to sternocleidomastoid muscle. Formed by convergence of retromandibular and posterior auricular veins. The external jugular drains into the subclavian vein lateral to the internal jugular. An anterior jugular vein system, may drain into the external or internal jugular.

G. Brachiocephalic vein. Two large veins formed by convergence of internal jugular and subclavian veins deep to the sternoclavicular joint.

H. Lymphatic drainage from head and neck. Lymphatics tend to follow arteries and veins. There is a ring of superficial nodes and a ring of deep nodes around the head and neck. These nodes usually have regional names and superficial nodes generally drain into deep nodes. Efferent channels from nodes converge to form vessels which drain into the subclavian veins on the left (thoracic duct) and right (right lymphatic duct) sides.

I. Thoracic duct. A major lymphatic channel that begins in the abdomen, ascends through the thorax between the descending aorta and azygous vein, and empties into the left subclavian vein, usually lateral to the point where the internal jugular joins the subclavian vein. The thoracic duct collects lymph from all of the body except the right half of the body superior to the diaphragm and returns the lymph to the venous system. (Lymph from the right half of the body superior to the diaphragm returns to the venous system by the right lymphatic duct.)

XIII. Pharynx. The funnel-shaped area common to the cranial end of the respiratory and alimentary passages. The pharynx begins posterior to the nose, at the base of the skull and descends to the cricoid cartilage. At the cricoid cartilage, the pharynx divides into the anteriorly placed larynx and the posteriorly placed esophagus. Three parts of pharynx: nasopharynx, posterior to the nose; oropharynx, posterior to the oral cavity; and laryngopharynx, posterior to the larynx. The pharyngeal isthmus joins the nasopharynx to the oropharynx. The isthmus is closed during swallowing by the soft palate elevating posteriorly to meet the posterior wall of the pharynx. This prevents food and liquids from passing through the nose. During respiration, the soft palate acts as a flap valve to allow continuity between the naso- and oropharynx.

A. Nasopharynx. Superior to soft palate. The auditory (eustachian) tube, with its cartilaginous extension, the torus tubarius, opens into this area. Deep to the mucous membrane, anterior to the auditory tube, is the tensor veli palatini and posterior to the tube, is the levator veli palatini. Extending inferiorly from the torus tubarius is the salpingopharyngeal fold, formed by the underlying salpingopharyngeus (tubal part of palatopharyngeus) muscle. Lymphatic tissue adjacent to the opening of the auditory tube constitutes the tubal tonsil. Lymphatic tissue on the posterior superior aspect of the nasopharynx constitutes the pharyngeal tonsil (= adenoids when enlarged).

B. Oropharynx. Inferior to soft palate and posterior to oral cavity. Palatoglossal arch (fold), formed by underlying muscle of the same name, separates oral cavity from oropharynx. Palatopharyngeal arch (fold) is also formed by underlying muscle of same name. Between the palatoglossal and palatopharyngeus arches is a triangular depression, the fauces, which contains the palatine tonsils. The inferior portion of the anterior wall of the oropharynx is formed by the root of the tongue and the epiglottic cartilage. The collection of lymphoid tissue at the root of the tongue constitutes the lingual tonsil. Three mucous membrane folds, consisting of one median glossoepiglottic fold between the tongue and epiglottic cartilage, and two lateral glossoepiglottic folds, between the epiglottis and the junction of the tongue and pharynx, bound depressions, the epiglottic valleculae.

C. Laryngopharynx. Posterior to larynx and extending from the inlet of the larynx inferiorly to the cricoid cartilage, where it becomes the esophagus. An important feature of this portion of the pharynx is the piriform recess, on each side of the laryngopharynx. Bones in a bolus of food may penetrate the mucosa and become stuck in this recess.

D. Layers of pharyngeal wall, from external to internal.
   1. Outermost layer=bucopharyngeal fascia, attached superiorly to pterygomandibular raphe (from pterygoid hamulus to mandible).
   2. Muscular layer. Outer circular and inner longitudinal layers. Three muscles, the superior, middle, and inferior constrictors stack on one another, like flower pots, and comprise the outer circular layer. Gaps between or above these muscles transmit important structures entering or leaving the pharynx: The gap cranial to the superior constrictor transmits the auditory tube and levator palati. The gap between the superior

and middle constrictors transmits nerves and vessels to the tongue and styloglossus and stylopharyngeus muscles. The gap between the middle and inferior constrictors transmits the superior laryngeal artery and internal laryngeal nerve. Inferiorly, between the inferior constrictor and esophagus, the inferior laryngeal nerve and vessels ascend to enter the larynx. The inner longitudinal layer is comprised of three muscles: salpingopharyngeus, palatopharyngeus, and stylopharyngeus.

    3.   Strong submucosa called pharyngobasilar fascia.

    4.   Mucosa.

  E.   Arteries. Include ascending pharyngeal (from external carotid) and pharyngeal branches of facial and maxillary arteries.

  F.   Lymphatics. Lymph drains into retropharyngeal and deep cervical nodes.

  G.   Nerves. Pharyngeal plexus, consists of IX, X, and sympathetic components. All muscles except tensor palati and stylopharyngeus are innervated by X. Tensor palati is innervated by V^3 and the stylopharyngeus by IX. The IX nerve is an important sensory nerve of the pharynx. In addition to the posterior one-third of the tongue, the nerve is sensory to the lateral aspect of the pharynx, in the region between the palatoglossal and palatopharyngeal arches. Stimulation of the mucous membrane in these areas produces a gag reflex and is a specific test for the sensory component of IX. The sensory innervation of the soft palate is V^2 and the sensory innervation of the pharyngeal mucosa posterior or inferior to the palatopharyngeal arch is X. Thus, stimulation of the soft palate or dorsal pharyngeal wall might also produce a gag reflex, but the effect would not be specific for IX.

XIV. Larynx. The "watchdog of the airway," lies in the neck between fourth and sixth cervical vertebrae. The larynx has a cartilaginous skeleton to maintain its patency for air flow, but the air flow may be decreased or cut off completely by voluntary muscles controlling the vocal cords. In addition to regulating the airway, the larynx functions as a voice production organ.

  A.   Cartilaginous skeleton: Five major cartilages of the larynx: cricoid, shaped like a signet ring, is most inferiorly placed and lies at the level of C6, with its lamina facing posteriorly; thyroid, forms the prominence of the Adam's apple; epiglottis, leaf-shaped and its superior end is free while the lateral margins are enclosed in the aryepiglottic folds; arytenoids (2 of these) are three-sided pyramidal structures sitting on the upper border of the cricoid lamina. Minor cartilages include the paired corniculates, at the apex of the arytenoid cartilages; the cuneiform cartilages, in the aryepiglottic folds and the triticeal cartilages, embedded in the free posterior edges of the thyrohyoid membrane.

  B.   Spaces and folds: The larynx is divided into three spaces or regions by the superior vestibular (false vocal cords) and inferior ventricular (true vocal cords) folds. The superior space, extending from the aditus to the vestibular folds, is the vestibule. The middle space is the ventricle, bounded by the vestibular folds superiorly and the ventricular folds inferiorly. Between the vestibular folds=rima vestibuli; between the ventricular folds= rima glottidis. The most inferior space, inferior to the rima glottidis= infraglottic portion, which leads into the trachea.

  C.   Muscles.

    1.   Sphincter muscles= sphincteric function at the laryngeal aditus and vestibule and help prevent food and water from entering the larynx.

        a)   Oblique arytenoid: Thin crossed muscles which overlie the transverse arytenoid muscle.

b) Aryepiglottis: Continuation of oblique arytenoid into aryepiglottic fold.

c) Thyroepiglottis: On medial side of quadrangular ligament, may help pull down the epiglottis during swallowing.

2. Muscles which control opening and closing of airway at rima glottidis:

a) Posterior cricoarytenoid. The only abductor of the vocal folds. Opens airway during respiration.

b) Lateral cricoarytenoid. Adducts or approximates the vocal folds.

c) Arytenoideus. Powerful adductor of vocal folds.

3. Muscles which regulate tension on the vocal folds:

a) Cricothyroid: anteriorly placed and upon contraction tilts the thyroid cartilage anteriorly at cricothyroid joint, thus increasing tension on true vocal ligaments.

b) Thyroarytenoid: Parallels vocal folds and lies lateral to them. Pulls arytenoids anteriorly, with consequent shortening and a decrease of tension on the vocal ligaments.

c) Vocalis: Consists of innermost muscle fibers of thyroarytenoid, attaching to vocal cords. Minutely adjusts vocal cords for speaking and singing.

D. Arteries. Laryngeal branches of superior and inferior thyroid arteries supply the larynx.

E. Lymphatics. Lymphatic drainage superior to true vocal folds is cranial to superior deep cervical nodes near the carotid bifurcation. Lymphatic drainage inferior to true vocal folds is caudally to inferior deep cervical nodes.

F. Nerves. Larynx innervated by superior and inferior laryngeal branches of X. The internal branch of superior laryngeal supplies sensory fibers to laryngeal mucosa superior to true vocal folds. The external branch of superior laryngeal innervates one muscle, the cricothyroid. The inferior laryngeal nerve, the continuation of the recurrent laryngeal nerve, supplies sensory fibers to the laryngeal mucosa inferior to the true vocal folds and motor fibers to the remaining intrinsic laryngeal muscles.

XV. Potpourri of radiological considerations.

A. Orbital shadows: Enlarged on AP view.

B. Pineal gland: Should be in midline and may be calcified.

C. Paranasal sinuses: Always viewed from both PA and lateral projections.

D. Sella turcica (of sphenoid): Enlargement signals abnormality or disease process.

E. Dura and choroid plexus: May have some calcification.

F. Optic canal (sphenoid): Seen end on end with oblique projection.

G. Orbit floor: Formed by roof of maxillary sinus.

H. Condylar process (mandible): Frequently fractured part of mandible.

I. Internal auditory canal: Viewed with PA projection.

J. Parotid and submandibular ducts: Visualized after injection of contrast material into duct orifices=sialogram.

K. Thyroid gland: Visualized after oral administration of radioactive iodine; viewed via nuclear scan.

L. Left vertebral artery: Usually larger than right.

M. Middle meningeal artery: Enlarged in certain brain tumors.

N. Internal carotid artery: No branches in neck, larger and posterior to external.

XVI. Applied Anatomy/Anatomical Pearls.

A. Facial nerve: Inflammation in the confined space of facial canal exerts pressure on the nerve; may cause ipsilateral paralysis in facial muscles=Bell's Palsy.

B. Facial nerve: Lesion of the lingual nerve (branch of trigeminal) would only alter general sensation to the tongue. Lesion of lingual nerve distal to the point where chorda tympani joins, would in addition, block taste to anterior 2/3s of tongue and secretion of submandibular and sublingual glands.

C. Helpful generalizations regarding cranial nerves:
    1. III, IV, VI and XII: Contain GSE components and innervate muscles that do not develop in association with branchial apparatus.
    2. V, VII, IX and X: Contain SVE components and innervate muscles that develop in association with branchial arches.
    3. III, VII, IX and X: Contain parasympathetic GVE components to viscera.
    4. V, VII, IX and X: Contain sensory components and have sensory ganglia homologous to dorsal root ganglia of spinal nerves.
    5. Only II and VIII contain SSA components.
    6. Taste (SVA): Carried over cranial nerves VII, IX, and X.
    7. Trigeminal nerve: Each major branch of this nerve has a parasympathetic ganglion attached to it: Ciliary ganglion, functionally related to III, is attached to V^1; pterygopalatine ganglion, functionally related to VII, is attached to V^2; submandibular ganglion, also functionally related to VII, is attached to V^3; and otic ganglion, functionally related to IX, is attached to V^3.
    8. Left recurrent laryngeal nerve: An enlarged heart may exert pressure on this nerve resulting in a hoarse voice.
    9. Oculomotor nerve: Emerges from brain stem between posterior cerebral artery and superior cerebellar artery; vulnerable to compression and dysfunction by aneurysms in either of these vessels.

D. Referred pain: Pain may originate from the diaphragm (phrenic, C_{3-5}) and be referred to skin over the clavicle (supraclavicular nerves $C_{3,4}$).

E. Middle meningeal artery: Epidural, may produce an epidural hemorrhage if injured.

F. Cavernous sinus: The following pass through or are embedded in the wall of this structure: Cranial nerves III, IV, V^1, V^2, VI, internal carotid artery.

G. Pterygoid plexus of veins: Deep in face, connect surface veins, such as the facial, with the cavernous dural venous sinus.

H. Superior laryngeal nerve: Injury to this nerve will impair or abolish the cough reflex.

I. Larynx: If surgically removed, coughing or lifting heavy objects is impossible, because the airway cannot be closed.

J. Piriform recess: Foreign bodies, such as bones, frequently lodge in this space in the laryngopharynx.

K. Trigeminal neuralgia (tic douloureux): Excruciating pain along one or more branches of the trigeminal nerve. The exact cause is unknown but the condition is often associated with an anomalous vessel lying adjacent to the trigeminal ganglion.

THORAX

I. Surface Anatomy.

   A. Nipple: An important landmark on the anterior thoracic wall, located at the fourth interspace, in the midclavicular line.

   B. Diaphragm: The right hemi-diaphragm rises to less than half an inch below the right nipple; the left hemi-diaphragm is about an inch below the left nipple. The central tendon is at the level of the xiphisternal junction. Posteriorly, the xiphisternal junction occurs at the level of the intervertebral disk between T9 and T10.

   C. Projection of heart: Right border parallels the right side of the sternum, extending lateral to the sternum less than an inch and extending vertically from the third to sixth costal cartilage. Lower border extends across xiphisternal junction toward the left to the midclavicular line at the fifth interspace (apex beat); Left border extends from apex beat to left 2nd intercostal space, about an inch from the left sternal margin. Superiorly placed base of heart extends from left second intercostal space to right superior heart border at third costal cartilage.

   D. Auscultation of heart valves: Auscultation of individual valves is dependent on the thickness of the chest wall and direction of blood flow. The pulmonic valve is auscultated at the left second intercostal space, near the sternal margin; the aortic valve is auscultated at the right second intercostal space, near the sternal margin; the mitral valve is auscultated at the apex beat (midclavicular line at the left fifth intercostal space), and the tricuspid is auscultated at the right margin of the sternum, at the fifth intercostal space.

II. Skeleton. Components of the skeleton of the thorax include the sternum, with its manubrium, body and xiphoid process, the ribs and costal cartilages, and thoracic vertebrae. The sternal angle (of Louis) is an important landmark on the anterior thoracic wall and marks the site where the second rib articulates at the junction of the manubrium and body of the sternum. Ribs 1-10 articulate with the sternum via costal cartilages. Ribs 11 and 12 are short, and fail to reach the sternum. Instead they terminate in muscle posteriorly. The thoracic inlet is relatively small and kidney-shaped. The thoracic outlet is large and irregular and closed by the diaphragm.

III. Muscles of thoracic wall: The muscular component of the thoracic wall is segmental and innervated by ventral primary rami of spinal nerves.

   A. Superficial layer: Muscles include the external intercostal and levatores costarum. External intercostal fibers, directed inferiorly and medially, extend from tubercles of ribs to costochondral junctions. At this junction, the muscle extends medially as the external intercostal membrane. Levatores costarum are deeply placed posteriorly. They arise from transverse processes of vertebrae and pass inferiorly to insert on the next lower rib.

   B. Intermediate layer: Internal intercostals comprise this layer. Muscle fibers, directed posterolaterally, extend from the angles of the ribs to the sternum.

   C. Innermost layer: Represented by 3 muscles: The innermost intercostals, which lie in one intercostal space, and the transversus thoracis, anteriorly, and subcostalis, posteriorly. The latter two muscles span more than one intercostal space in their respective locations.

IV. Intercostal vessels and nerves: These vessels and nerves run between the intermediate and innermost muscular layers of the thoracic wall. Their arrangement in each intercostal space from superior to inferior is: vein, artery, and nerve. The mnemonic is VAN.

A. Posterior intercostal arteries. There are 11 pairs of these. The cranial two pairs usually arise from the costocervical trunk and the remainder arise from the thoracic aorta.

B. Anterior intercostal arteries. These arise from the internal thoracic artery, perforate through the thoracic wall lateral to the sternum, and anastomose with the posterior intercostals.

C. Intercostal veins: Anterior veins drain into internal thoracic veins and posterior veins drain into the azygos vein on the right side and the hemiazygos system on the left side.

V. Abdominal diaphragm: The large muscle is attached to the sternum, costal cartilages and ribs, and posteriorly, to vertebrae. The posterior attachment constitutes the lumbar portion and two crura are found here. These crura form the sides of the aortic hiatus through which the aorta and thoracic duct pass at the T12 vertebral level. The crura are connected by connective tissue, the median arcuate ligament. On either side of the median arcuate ligament, are medial arcuate ligaments (medial lumbocostal arch), which are thickenings of fascia over the cranial aspect of the psoas major. Lateral and inferior to the medial arcuate ligaments, are the lateral arcuate ligaments (lateral lumbocostal arch). These are thickenings of fascia over the quadratus lumborum muscle.

A. Innervation: motor and sensory from the phrenic nerve (C_{3-5}) off the cervical plexus.

B. Vascular supply: Inferior phrenics (aorta), musculophrenic (internal thoracic) and pericardiacophrenic (internal thoracic).

C. Structures passing through: Inferior vena cava at T8 vertebral level via the caval foramen, esophagus (and vagal trunks) at T10 vertebral level via the esophageal hiatus, and the aorta (and thoracic duct) at T12 vertebral level via the aortic hiatus.

VI. Pleura: This is a thin serous membrane, invaginated by the ingrowing lung bud during development, thus forming the definitive parietal and visceral pleura and the pleural cavity.

A. Parietal. The parietal pleura is adherent to the inner aspect of the thoracic wall like wall paper to a wall. Four parts of parietal pleura include: mediastinal, lining the mediastinum; diaphragmatic, lining the diaphragm; costal, lining the thoracic wall; and superiorly, the cupola, which extends superior to the first rib. The line where costal pleura becomes diaphragmatic pleura is called the costodiaphragmatic reflection. Similarly, the line where costal pleura becomes mediastinal pleura is called the costomediastinal reflection. In cases of thoracic injury, the lines of pleural reflection determine whether the injury involves the pleura cavity, therefore the surface projection of theses lines of reflection is important. The right pleural reflection passes across the sternoclavicular joint and proceeds (near the middle of the sternum) inferiorly from the level of the second rib to the 6th costal cartilage, swings laterally to cross the 8th rib in the midclavicular line, the 10th rib in the midaxillary line, and the 12th rib posteriorly, near the midline. The mnemonic for the right side is 2, 6, 8, 10, 12. The left pleural reflection is similar, except at the 4th costal cartilage the line swings laterally to the left border of the sternum. The mnemonic is 2, 4, 6, 8, 10, 12 for the left side.. At the right and left costodiaphragmatic reflection, pleura extends caudally without intervening lung tissue. This caudal extension forms a potential space (in the pleural cavity), the costodiaphragmatic recess. Similarly, at the costomediastinal reflection on the left side, because of the cardiac notch in the left lung, lung tissue does not extend up to the costomediastinal reflection and another potential space, the costomediastinal recess, is formed. Parietal pleura has a rich sensory innervation from the phrenic and intercostal nerves.

B. Visceral. This innermost layer is applied to and inseparable from the substance of the lung. It is continuous with parietal pleura at the hilus of the lung.

C. Pleural cavity. The potential space on each side of the thoracic cavity between the visceral and parietal pleura. Normally this space has only a thin film of serous lubrication.

VII. Lungs. The trachea, lying anterior to the esophagus, leads to the lungs and at the sternal angle it divides into right and left primary bronchi. Within the trachea, this division is marked by the raised carina. The right primary bronchus is shorter, more vertical and a larger diameter than the left and objects aspirated into the airway are more likely to become lodged in the right primary bronchus. Usually there are two left and one right bronchial arteries arising from the aorta or posterior intercostal arteries.

A. Right lung: Divided into 3 lobes: upper, middle, and lower by horizontal (superiorly) and oblique fissures. The horizontal fissure, separating the upper and middle lobes, runs from the oblique fissure and follows the fourth costal cartilage toward the sternum. The larger oblique fissure, tends to follow the 6th costal cartilage anteriorly. Ten bronchopulmonary segments, named by their anatomical position, are found in the right lung.

B. Left lung: Divided into 2 lobes: upper and lower, by well developed oblique fissure. The lingula of the upper lobe is believed to be homologous to the middle lobe of the right lung.

VIII. Mediastinum: The space in the thorax between the two pleural cavities which houses the heart and numerous other structures. For purposes of description, it is divided into various portions by the placement of an imaginary line and the position of the heart. An imaginary line (horizontal plane) extending from the sternal angle posteriorly intercepts the intervertebral disk between T4 and T5 and divides the mediastinum into superior and inferior portions. The inferior mediastinum is further subdivided into anterior, middle (occupied by heart), and posterior parts.

A. Superior: This space contains structures entering and leaving the inferior mediastinum including brachiocephalic veins and superior vena cava, arch of aorta and beginning of its 3 great vessels, vagus and phrenic nerves, trachea, esophagus and thoracic duct. The thymus is also here.

B. Anterior: This space is anterior to the pericardial sac and contains fat, connective tissue, and the internal thoracic vessels.

C. Middle: Contains the heart and pericardial cavity. The phrenic nerves pass between the parietal pericardium and mediastinal pleura.

D. Posterior: This space is posterior to the heart, between the left and right mediastinal pleura. Contents include the descending aorta, esophagus, azygos venous system, thoracic duct and vagus nerve derivatives (esophageal plexus, vagal trunks).

IX. Pericardium. The serous sac that surrounds the heart. Similar to pleura, developmentally, in that the heart grew into the sac and part of the sac, the visceral pericardium, becomes adherent to and inseparable from the heart, and is also called the epicardium. Visceral pericardium is continuous with parietal pericardium at the base of the heart and at the roots of the great vessels entering and leaving the heart. The parietal pericardium is reinforced with tough fibers externally.

The space between the visceral and parietal pericardium, occupied by the heart, is the pericardial cavity.

X.   Heart.

  A.   Surface features.
    1.   Sulci: There are three of these which are usually covered with fat. The coronary sulcus, separates atria from ventricles and the anterior and posterior interventricular sulcus separates the ventricles anteriorly and posteriorly.
    2.   Coronary arteries: A right and left coronary artery arises from the ascending aorta at the right and left aortic sinus.
      a)   Right: Branches of the right coronary artery include the anterior right atrial branch (ARAB) which usually supplies an important branch to the SA node; a right marginal branch, which supplies the right ventricle; and the posterior interventricular branch, running in the posterior interventricular sulcus, supplying both ventricles.
      b)   Left: Divides shortly after its origin into anterior interventricular, which descends in sulcus of same name and is a major supply to the left ventricle; and circumflex, which runs posteriorly in the coronary sulcus to anastomose with the right coronary. The circumflex may give off (left) marginal branches to the left ventricle.
    3.   Cardiac veins: Drain the heart muscle.
      a)   Great: Parallels the anterior interventricular artery, and goes posteriorly in the coronary sulcus, and after receiving the oblique vein of the left atrium, becomes the coronary sinus, which empties into the right atrium.
      b)   Middle: Parallels the posterior interventricular artery and drains into the coronary sinus.
      c)   Small: A variable vein that drains the right atrium primarily and empties into the coronary sinus nears its termination.
      d)   Anterior: Arise on the anterior surface of the right ventricle, cross the coronary sulcus, and empty directly into the right atrium.

  B.   Right atrium, topographical features: The posterior smooth walled portion is the sinus venarum; the anterior muscular portion contains the pectinate muscles. The crista terminalis is a muscular ridge separating these two portions. The superior vena cava, inferior vena cava, and coronary sinus open into this chamber. The posterior interatrial area contains a depression, the fossa ovalis, and the ridge bordering the fossa is called the limbus fossa ovalis. The right auricle is the superiorly placed evaginated pouch extending from the muscular portion of the right atrium.

  C.   Right ventricle, topographical features: Contains muscular fasciculi, the trabeculae carnae, except for a small, smooth walled portion, the conus arteriosus, inferior to the pulmonary trunk orifice. The tricuspid valve surrounds the atrioventricular orifice. Usually there is a large anterior papillary muscle and a smaller posterior papillary muscle. Running from papillary muscles to the free edge of the three fibrous valve cusps are tough fibrous cords, the chordae tendineae. These cords prevent valve cusps from everting during systole. Chordae tendineae attaching to the septal cusp usually arise directly from the interventricular septum while chordae tendineae attaching to the anterior and posterior cusps usually arise from anterior and posterior papillary muscles. The septomarginal trabeculae (moderator band) extends from the septum to the anterior papillary muscle and carries with it the right branch of the atrioventricular bundle. The pulmonary trunk, placed superiorly, is the exit from this chamber.

D. Left atrium, topographical features: The left atrium is posterior in the middle mediastinum and receives the 4 pulmonary veins. Like the right atrium, it has a large smooth portion and a smaller muscular portion (pectinate muscles), with an auricle. A depression on the interatrial septum marks the fossa ovalis.

E. Left ventricle, topographical features: Forms the apex of the heart. Its wall is three times thicker than the right ventricular wall. Like the right ventricle, it has trabeculae carneae, and the structure of the left atrioventricular valve is similar to the right except there are only two valve cusps=bicuspid or mitral valve, and two large papillary muscles, an anterior and posterior. The aorta opens superiorly from the left ventricle.

F. Conducting system: The heart muscle has an intrinsic contractile auto-rhythmicity, which is initiated at the sinuatrial (SA) node. This node is located in the anterior wall of the right atrium, near the junction of the superior vena cava. The wave of depolarization spreads over the atria and is picked up by the atrioventricular (AV) node, located in the interatrial septum, superior to the opening of the coronary sinus. Extending from the AV node down the interventricular septum, are right and left bundle branches (of His) which carry the electrical impulse into the ventricles. All conductive tissue (nodes, bundle of His) is specialized cardiac muscle.

G. Aorta. Ascends from the left ventricle. Near its beginning, the 3 cusps of the aortic semilunar valves can be located. The area superior and lateral to each cusp is an aortic sinus. Right and left coronary arteries arise from the right and left aortic sinus. The ascending aorta arches behind the manubrium and gives off the brachiocephalic, left common carotid, and left subclavian arteries. On the concave side of the aortic arch, the fibrous ligamentum arteriosum connects to the pulmonary trunk.

H. Pulmonary trunk. Ascends from the right ventricle and divides into right and left branches. The right branch goes posterior to the arch of aorta to reach the right lung and the left branch goes anterior to the proximal descending aorta to reach the left lung. The pulmonary semilunar valve, similar to the aortic semilunar valve, is located near the beginning of this vessel.

XI. Autonomic nervous system in thorax.

A. Parasympathetics: 2-3 preganglionic cardiac branches from cervical and thoracic portion of vagus descend to cardiac plexus.

B. Sympathetics: 2-3 sympathetic postganglionic cardiac branches from right and left cervical sympathetic trunks descend to cardiac plexus. Sympathetic branches may travel with vagal cardiac branches.

C. Cardiac plexus: Ganglionated plexus, located in concavity of arch of aorta, continuous over tracheal bifurcation and bifurcation of pulmonary trunk. Contains postganglionic parasympathetic nerve cell bodies, and parasympathetic, sympathetic, and visceral sensory fibers.

D. Pulmonary plexus: Continuation of cardiac plexus onto tracheal bifurcation and root of lung. Vagal, sympathetic, and GVA components.

E.   Esophageal plexus: Primarily right and left vagal components that emerge from or never enter into pulmonary plexus. Also contains sympathetic components (from greater splanchnic nerve) and sensory components. This plexus innervates the esophagus, but in a quantitative sense, most fibers continue caudally, pass through the esophageal hiatus, and enter the abdomen as anterior and posterior vagal trunks.

F.   Thoracic splanchnic nerves: The greater splanchnic nerves are derived from T_{5-9} segmental levels; the lesser from $T_{10,11}$; the least from T_{12}. Splanchnic nerves are primarily preganglionic components that pass through the posterior aspect of the diaphragm and synapse in prevertebral ganglia. The greater splanchnic synapses primarily in the celiac ganglion, the lesser splanchnic synapses primarily in the aorticorenal ganglion, and the least splanchnic synapses primarily in the renal plexus. Visceral sensory fibers accompany sympathetic components in splanchnic nerves.

XII.   Potpourri of radiological considerations.

A.   Heart, PA projection: right heart border=right atrium, left heart border=left ventricle, right ventricle and left atrium not seen.

B.   Heart, left lateral projection: anterior border=right ventricle, posterior border=left atrium and left ventricle, inferior shadow from below=inferior vena cava.

C.   Heart, AP projection: Left lower heart border= left ventricle, right lower heart border=right atrium.

D.   Thymus: widens mediastinum in infants and young patients.

XIII.   Applied anatomy/anatomical pearls.

A.   Trachea: bifurcates at T4.

B.   Lungs and pleural cavity: extend cranially into neck, superior to the clavicle.

C.   Referred pain from heart: Felt in body wall innervated by upper 5 thoracic spinal nerves, usually on left side.

D.   Sternal angle: Reference point for accurately identifying rib number or intercostal space.

E.   Ribs: Usually fractured at the angle, the weakest part.

F.   Thoracocentesis: Needle placed over superior border of rib to avoid damage to intercostal vessels and nerve.

G.   Deceleration injuries: May shear great vessels from arch of aorta.

H.   Right primary bronchus: Larger and more vertical than left. Objects are usually aspirated into this bronchus.

I.   Pericardial cavity: Excess fluid here may compress heart and lead to death (cardiac tamponade).

ABDOMEN

I.   Ventrolateral abdominal wall.

A.   Surface anatomy.

1.   Iliac crest: an important surface landmark, posteriorly. A line connecting the 2 crests crosses the L4 vertebrae. This line is a useful landmark for lumbar punctures since spinal cord in adult does not extend inferior to L2 vertebral level.
2.   Anterior superior iliac spine: palpable, inguinal ligament attaches here.
3.   Umbilicus: At level of intervertebral disk between L3 and L4.
4.   Semilunar line: Curved lateral margin of rectus abdominis extending from pubic tubercle to 5th costochondral junction.

B.   Superficial fascia. In the abdomen, this fascia covering the anterolateral abdominal muscles can usually be separated into an outer fatty layer (Camper's) and a deeper membranous layer (Scarpa's). This lamination is more obvious about midway between the umbilicus and pubic crest. The attachments of Scarpa's fascia are important because these attachments determine, in cases of trauma to the lower abdomen/pelvis, where fluids (blood, urine) accumulate subcutaneously. The attachments of Scarpa's fascia are: along the iliac crest, inferiorly to fascia lata, paralleling the inguinal ligament, and medially to the pubic symphysis. Scarpa's fascia extends into the perineum (as Colles' fascia), attaching laterally to the ischiopubic rami and posteriorly to the base of the urogenital diaphragm. Cutaneous nerves are found in the superficial fascia: For reference, T_{10} is at level of umbilicus; L_1 is at level of pubic crest.

C.   Muscles of the anterolateral abdominal wall. All are segmentally innervated and vascularized.
1.   External abdominal oblique. Outermost muscle whose fibers are directed inferiorly and medially. Has a broad aponeurotic portion anteriorly and inferiorly that forms several important structures:
   a)   Inguinal ligament: rolled under inferior border of external abdominal oblique extending from anterior superior iliac spine to pubic tubercle.
   b)   Lacunar ligament: curved medial fibers of inguinal ligament which pass posteriorly and turn laterally on the pecten pubis. The lateral continuation of these fibers=pectineal ligament.
   c)   Superficial inguinal ring: An opening in the aponeurosis over the pubic tubercle, transmitting the spermatic cord. The upper side of this opening=medial (superior) crus, lower side of this opening=lateral (inferior) crus. Intercrural fibers help prevent enlargement of the superficial ring.
2.   Internal abdominal oblique. Deep to external oblique. Fibers run at a right angle to external abdominal oblique, an arrangement which lends strength to the abdominal wall. Fibers of the internal oblique which originate from the lateral half of the inguinal ligament become aponeurotic medially and arch medially for attachment to the pubic tubercle. These arching fibers contribute to the formation of the conjoint tendon (falx inguinalis) along with similar contributions from the transversus abdominis.
3.   Transversus abdominis. The innermost of the abdominal flank muscles, with transversely oriented muscle fibers. Aponeurotic components from this muscle join similar components from the internal oblique to form the conjoint tendon (falx inguinalis). Layers of the abdominal wall deep to the transversus abdominis are: transversalis fascia, extra-peritoneal fat and connective tissue and the peritoneum.
4.   Rectus abdominis. Vertically oriented muscle running from the pubis to the rib cage.
   a)   Rectus sheath: The aponeurosis of the three flank muscles separate or fuse at the lateral edge of the rectus abdominis forming a sheath for the muscle. The formation and composition of this sheath changes at different levels. Vertically, two to three inches on either side of the umbilicus, the aponeurosis of internal

oblique splits into anterior and posterior components at the lateral edge of the muscle. Anterior components join aponeurotic components of external oblique to form the rectus sheath anteriorly. A posterior component from the internal oblique joins the aponeurosis of transversus abdominis to complete the rectus sheath posteriorly. About midway between the umbilicus and pubis, the internal oblique aponeurosis no longer splits at the lateral margin of rectus abdominis. Instead, the aponeurotic components of external oblique, internal oblique, and transversus abdominis pass anterior to rectus abdominis. Thus, posteriorly, the rectus abdominis lies against transversalis fascia. The point where this transition in sheath components occurs is seen superficially as the arcuate line.

    b)   Inferior epigastric artery. A branch of the external iliac artery. Originates medial to deep inguinal ring and enters rectus sheath to lie on posterior aspect of rectus abdominis. Anastomoses with superior epigastric artery, from internal thoracic. This anastomosis constitutes a potentially important collateral route between the subclavian and external iliac arteries.

D.   Inguinal canal. This canal transmits the spermatic cord in males and the round ligament of the uterus in females. Entrance into the canal is via the deep inguinal ring, a finger-like diverticulum of transversalis fascia, just lateral to the inferior epigastric artery. Exit from the canal is via the superficial inguinal ring. The boundaries of the canal are: floor, inguinal and lacunar ligaments; roof, arching fibers of internal oblique and transversus abdominis; anterior wall, aponeurosis of external oblique; and posterior wall, transversalis fascia and conjoint tendon.

    1.   Direct inguinal hernia: Protrusion of an abdominal viscus through the inguinal triangle (between lateral border of rectus abdominis and inferior epigastric artery) into the superficial inguinal ring.

    2.   Indirect inguinal hernia: Protrusion of an abdominal viscus along the path of the descent of the testis: viz through the deep inguinal ring, and through the inguinal canal and superficial inguinal ring and often into the scrotum.

E.   Testis and spermatic cord. As the testis and spermatic cord pass through the inguinal canal, they obtain coverings derived from layers of the abdominal wall.

Layer of Abdominal Wall	Corresponding Layer
skin	skin of scrotum
superficial fascia	perineal fascia and dartos muscle
external oblique	external spermatic fascia
internal oblique	cremaster muscle
transversus abdominis	none
transversalis fascia	internal spermatic fascia
extraperitoneal fat	areolar tissue in spermatic cord
peritoneum	tunica vaginalis testis

II.   Peritoneum. The serous membrane that lines the abdominopelvic cavity.

A.   Parietal: Innermost lining of the abdominal wall which effectively forms a closed sac except in the female, where the closed sac communicates with the exterior via the uterine tubes, uterus and vagina.

B.   Visceral: The peritoneum that developed from splanchnic mesoderm and thus adhering to viscera is called visceral peritoneum or serosa.

C. Retroperitoneal structures: Some organs develop deep to parietal peritoneum. Usually these organs are covered by peritoneum on only one surface. Such organs are considered primarily retroperitoneal and examples include the kidneys and urinary bladder. In other cases, organs initially are covered by visceral peritoneum and suspended by a mesentery, but later in development, these organs are pushed to the sides of the peritoneal cavity and the peritoneum on the deep side of the organ is lost. The result is peritoneum occurs on only one side of the organ. Such organs are considered secondarily retroperitoneal and examples include the ascending and descending colon.

D. Mesentery: A double layer of peritoneum that suspends the intestine from the dorsal body wall. Nerves and vessels reach the intestine by traveling between the two layers of peritoneum.

E. Lesser omentum: A double layer of peritoneum connecting the stomach (lesser curvature side) to the liver.

F. Greater omentum: A double layer of peritoneum connecting the stomach (greater curvature side) to the transverse colon and other structures. Part of the greater omentum is redundant and folds back on itself (4 layers of peritoneum) and descends to cover abdominal contents, like an apron.

G. Peritoneal ligaments: Another name for double layers of peritoneum connecting abdominal viscera. Examples include the gastrophrenic and gastrolienal ligaments, which are portions of the greater omentum connecting the stomach to the diaphragm and spleen, respectively.

H. Greater sac: The major peritoneal cavity.

I. Lesser sac: An evagination of the greater sac which extends posterior to the stomach. Also called the omental bursa. The lesser sac communicates with the greater sac via the epiploic foramen (of Winslow).

J. Peritoneal gutters: Pathways along dorsal parietal peritoneum which allows movement of fluids etc. from one part of the peritoneal cavity to another. Right and left paracolic gutters are lateral to the ascending and descending colon respectively, and are open superiorly and inferiorly. The gutter to the right of the dorsal mesentery is closed superiorly and inferiorly. The gutter to the left of the dorsal mesentery is closed superiorly and open inferiorly.

III. Gut and appendages.

A. Stomach: Parts include the cardia, around the esophageal opening; fundus, portion superior to esophageal opening; body, inferior to fundus; and pyloric portion, to the right of the angular notch (acute indentation approximately 2/3s way down lesser curvature).

B. Duodenum: The first part of the small intestine, approximately 12 inches long, roughly "C" shaped and largely secondarily retroperitoneal. It is divided into 4 parts: first part, approximately 2 inches long, is mesenteric; second part, retroperitoneal, descends 4-5 inches and receives the common bile duct on its medial side; third part, also retroperitoneal, continues medially between the aorta and superior mesenteric artery; and the fourth part ascends about 2 inches to join the jejunum. This juncture is marked by a strong fibromuscular band, the suspensory ligament (of Treitz), which represents the cranial end of the dorsal mesentery.

C.   Jejunum: The upper two-fifths of the small intestine; resides in upper left quadrant of abdominal cavity. Larger diameter than the ileum and arterial vasa recta are longer than those to ileum. Suspended by mesentery.

D.   Ileum: Terminal three-fifths of small intestine. Smaller diameter, shorter vasa recta than jejunum. Suspended by mesentery.

E.   Large intestine: The ileum joins the large intestine at the ileocecal junction. The part of the digestive tract inferior to this junction is the cecum with its attached appendix. Superior to the junction the segments of large intestine are: ascending colon, transverse colon, descending colon, sigmoid colon, rectum and anal canal. Ascending and descending colon are secondarily retroperitoneal, transverse colon and sigmoid colon and the superior one-third of the rectum are mesonteric intraperitoneal. Large intestine is identified by tenia coli, appendices epiploicae, and sacculations called haustra.

F.   Liver: Divided into right and left lobes by the falciform ligament. Anteriorly, to the right of the falciform ligament, is the quadrate lobe, and posteriorly to the right of the falciform ligament is the caudate lobe. Vessels and nerves enter and leave the liver between the caudate and quadrate lobes. Superior features of the liver include the coronary ligaments and bare area. Bile leaves the liver via the right and left hepatic ducts which fuse to form the common hepatic duct. The cystic duct, from the gall bladder, then joins the common hepatic duct. Inferior to this juncture, the duct is called the common bile duct. The common bile duct empties into the medial side of the second part of the duodenum. The entrance of this duct into the duodenum is marked by the major duodenal papillae and sphincter of Oddi.

G.   Spleen: Located in left, superior, abdominal cavity, deep to ribs 9-11.

H.   Pancreas: Secondarily retroperitoneal, dorsal to the stomach. Consists of head, neck, body, and tail portions. The head sits in the concavity of the "C" of the duodenum. The uncinate process is a portion of the head that hooks dorsally around the superior mesenteric vessels. These vessels form a groove in the pancreas dorsally, delineating the neck. The body continues toward the left side and the tail often makes contact with the spleen. The pancreatic duct joins the common bile at its entrance into the duodenum or the pancreatic duct may empty separately into the second part of the duodenum.

IV. Abdominal Vasculature.

A.   Branches of Abdominal aorta.
     1.   Paired visceral branches: Includes renal (L1 vertebral level) and gonadal (L2 vertebral level) and middle suprarenal (if present) vessels. Superior suprarenal arteries are usually branches of inferior phrenic arteries and inferior suprarenal arteries are usually branches of renal arteries. If middle suprarenals are present, they are branches of the aorta.
     2.   Unpaired visceral branches: Includes the celiac (T12/L1 vertebral level), with left gastric, common hepatic and splenic branches; the superior mesenteric (L1 vertebral level), with inferior pancreaticoduodenal, intestinal, ileocolic, right colic, and middle colic branches; and the inferior mesenteric (L3 vertebral level), with left colic, sigmoid, and superior rectal branches.
     3.   Paired parietal branches: Includes inferior phrenics (T12 vertebral level) and 4 pairs of lumbar arteries (L1-L4 vertebral levels). The aorta bifurcates at L4 into common iliac arteries.

4. Unpaired parietal branch: The representative of this group is the median sacral artery, which arises on the dorsal aspect of the aorta at its bifurcation into common iliacs.

B. Arterial anastomoses: The superior pancreaticoduodenal (from the celiac system) and the inferior pancreaticoduodenal (from the superior mesenteric) anastomose in the head of the pancreas. The middle colic (from superior mesenteric) and left colic (from inferior mesenteric) anastomose along the mesenteric border of the transverse colon. The superior rectal (from inferior mesenteric) and middle rectal (from internal iliac) anastomose on the sides of the rectum.

C. Inferior vena cava: Formed by union of common iliac veins at L5 vertebral level. Ascends in abdomen to right of aorta. Pierces diaphragm at T8 and enters the right atrium of the heart. Tributaries include 4 pairs of lumbar veins, renal veins, right gonadal, right suprarenal, right inferior phrenic veins and the short hepatic veins posterior to the liver. The left renal vein crosses anterior to the aorta and receives the left gonadal and left suprarenal. The left inferior phrenic usually drains into the left suprarenal.

D. Hepatic portal vein: In general, the term "portal" indicates a vein interposed between two capillary beds. The hepatic portal primarily collects blood from capillaries of the intestines and conveys it to the capillaries (sinusoids) of the liver. There are no functional valves in the hepatic portal venous system. Usually the inferior mesenteric vein drains into the splenic vein and the splenic vein merges with the superior mesenteric vein to form the portal. Alternatively, all three veins may merge to form the portal. The portal vein enters the liver posterior to the proper hepatic artery and common bile duct.

E. Portal hypertension. In several places, tributaries of the caval system and portal system anastomose. If the liver is diseased, blood flow through the organ is impeded. In such cases, sites of anastomoses often dilate and become varicose, and may break and hemorrhage. Important portal-caval anastomoses which may dilate with portal hypertension include: the esophagus, where tributaries of the left gastric vein (portal system) join tributaries of the azygos system (caval system); the rectum, where tributaries of the superior rectal vein (portal system) join the middle and inferior rectal veins (caval system); around the umbilicus, where veins in the falciform ligament (portal system) join intercostal veins (caval system); and retroperitoneally, where tributaries of the splenic and colic veins (portal system) join branches of the left renal vein (caval system).

F. Abdominal lymphatics. Efferent flow from the lower limbs, para-aortic, and intestinal nodes drains into a dilation at L2, the cisterna chyli, which marks the beginning of the thoracic duct. The thoracic duct ascends through the abdomen via the aortic hiatus.

V. Posterior abdominal structures.

A. Kidney: Retroperitoneal, extending from T12 to L3. The right kidney is usually slightly lower (half to one inch) than the left kidney because the large right lobe of the liver stopped the cranial migration of the organ during development. The kidney is stabilized by perirenal fat and connective tissue. Within the kidney, the ureter expands to form the renal pelvis; draining into the pelvis are 2-3 major calyces. Minor calyces, draining renal pyramids, converge to form the major calyces. The left renal vein is longer than the right, since it crosses anterior to the aorta to empty into the vena cava. Also, the right renal artery is longer and passes posterior to the vena cava to reach the kidney.

B. Suprarenal gland: Located superior and adjacent to the kidney, on either side of the celiac artery.

C. Muscles: All segmentally innervated by ventral primary rami.
   1. Psoas major: Originates on either side of lumbar vertebrae and extends caudally, posterior to the inguinal ligament, to insert on the lesser trochanter. Segmentally innervated.
   2. Iliacus: Occupies the iliac fossa, extends caudally, and joins the psoas, forming the powerful iliopsoas muscle, which inserts on the lesser trochanter and functions as a powerful flexor of the hip.
   3. Quadratus lumborum: Lies on the posterior abdominal wall, lateral to psoas. Originates caudally, from the iliolumbar ligament, and ascends to insert on the 12th rib and upper lumbar transverse processes. Important in lateral bending of the abdomen.

D. Somatic nerves: All derived from ventral primary rami.
   1. Subcostal: T_{12} ventral primary ramus.
   2. Iliohypogastric: Branch of anterior division of L_1, sensory to suprapubic region.
   3. Ilioinguinal: Branch of anterior division of L_1, exits superficial inguinal ring lateral to spermatic cord and provides cutaneous supply to superior medial thigh and anterior scrotum (labium majus).
   4. Genitofemoral: Branch of anterior division of $L_{1,2}$, exits the anterior border of psoas major, genital branch enters inguinal canal via deep inguinal ring to provide motor innervation to cremaster muscle. Femoral branch, more medial, supplies cutaneous innervation to superior medial thigh.
   5. Lateral femoral cutaneous: Posterior division of $L_{2,3}$, passes posterior to inguinal ligament medial to anterior superior iliac spine, sensory to lateral superior, thigh.
   6. Femoral: Posterior division of $L_{2,3,4}$, wedged between psoas major and iliacus on posterior body wall, exits abdomen deep to inguinal ligament, provides motor and sensory supply to anterior compartment of thigh.
   7. Obturator: Anterior division of $L_{2,3,4}$, descends on medial side of psoas major, passes inferiorly via obturator foramen, provides motor and cutaneous branches to adductor compartment of thigh.
   8. Lumbo-sacral trunk: Formed from ventral primary rami of $L_{4,5}$. Lies deep on the ala of the sacrum. Descends to join the sacral plexus.

VI. Autonomic nervous system.

A. Sympathetic trunks: Caudal continuation of thoracic sympathetic trunk lying in a paravertebral position on the anteromedial aspect of psoas major. Gives off 4 lumbar splanchnic nerves (preganglionic); upper two go to intermesenteric plexus, lower two go to superior hypogastric plexus.

B. Thoracic splanchnic nerves: Recall that GVE preganglionic components of the greater, lesser and least splanchnic nerves synapse in the celiac ganglion, aorticorenal ganglion, and renal plexus, respectively.

C. Vagal trunks: Anterior vagal trunks provide preganglionic parasympathetic GVE input to the stomach, duodenum, and pancreas. The posterior vagal trunk enters the celiac ganglion and vagal components distribute via the prevertebral (preaortic) ganglionated plexus to abdominal viscera. The vagus provides parasympathetic innervation to the gastrointestinal tract to the level of the left colic flexure. Parasympathetics to the rest of the gut are provided by pelvic splanchnic nerves.

D. Prevertebral (preaortic) ganglionated plexus. The ganglionated network of autonomic fibers around the abdominal aorta and its major branches. This plexus continues caudally across the pelvic brim as hypogastric nerves which connect with the inferior hypogastric plexus on each side of the pelvis. The prevertebral plexus has sympathetic, parasympathetic, and GVA components.

VII. Potpourri of radiological considerations.

A. Plain film of abdomen: In addition to bone, soft tissue including the liver, spleen, psoas margins and renal shadows may be identified.

B. Stomach: Best identified when distended with radiodense contrast material such as barium sulfate.

C. Hiatal hernia: Indicated by stomach protruding superiorly above diaphragm.

D. Ligament of Treitz: Site of termination of duodenal loop and beginning of jejunum.

E. Jejunum: Upper left quadrant of abdomen, feathery appearance (because of numerous plicae circulares) with barium.

F. Ileum: Lower right quadrant of abdomen, barium appears thicker.

G. Colon: Opacified by retrograde introduction of barium sulfate.

H. Gall bladder: Opacified by oral administration of contrast agent secreted by liver and concentrated in bile. Also identified with ultrasound.

I. Kidneys: Seen best after intravenous administration of water soluble contrast material which is excreted by kidneys.

J. Renal developmental anomalies: Includes duplicated ureters, fusion of lower poles of kidney (horseshoe), failure of ascent (pelvic kidney).

K. Uterine tube patency: Confirmed by introduction of contrast into cervical os and subsequent bilateral spillage of contrast into peritoneal cavity.

VIII. Applied anatomy/anatomical pearls.

A. Scarpa's fascia continues into the perineum as Colle's fascia. Rupture of the male urethra may lead to urine accumulation into the potential space (superficial perineal) deep to Colle's fascia.

B. Duodenal ulcers: Occur most often in the first part of the duodenum (duodenal bulb).

C. Volvulus: Twisting of a freely mobile portion of gut, often the sigmoid colon. Potentially dangerous if blood supply is compromised.

D. Megacolon (Hirschsprung's Disease): Distension of large intestine caused by failure of ganglia to develop in Auerbach's and Meissner's plexuses. Lack of ganglia results in atonia and loss of peristalsis.

E. Inferior vena cava: May transmit emboli from lower limb to right heart and subsequently to lungs.

F. Sympathetic nerves: Also carry pain fibers from abdominal viscera.

G. Right renal artery: Longer than left and lies posterior to the inferior vena cava.

H.	Left renal vein: Longer than the right and lies anterior to the aorta, immediately inferior to the superior mesenteric artery.

I.	Perirenal fat and fascia: Deficient inferiorly and may allow kidney to displace caudally (nephroptosis).

J.	Ureteric constrictions: Three of these: (1) at the pelvic-ureteric junction, (2) where ureter crosses pelvic brim, and (3) where ureters enter bladder. These are sites where renal stones often lodge during a caudal descent and cause painful reflex contractions of ureteric smooth muscle. GVA fibers from T_{11}-L_2 innervate the ureter and may refer pain to surface dermatomes.

K.	Parietal peritoneum: Innervated by GSA components of somatic nerves and pain emanating from this serous membrane can usually be precisely localized. Visceral peritoneum, however, is innervated by GVA components and painful irritation from this layer is poorly localized.

PELVIS AND PERINEUM

I.	Overview of pelvis. The pelvis is the inferior portion of the abdominopelvic cavity which houses the true pelvic viscera. The term pelvis is also used to indicate the bony surrounding of the inferior extent of the abdominopelvic cavity. The inferior extent of the pelvic cavity is closed by the pelvic diaphragm.

A.	Hip bone (os coxae). Each hip bone is formed by fusion of the pubis, ileum, and ischium. The two hip bones on each side plus the intervening sacrum form the bony pelvis.
1.	Articulations: The lumbosacral articulation, between L5 and S1, is stabilized by the iliolumbar ligament, running from the crest of the ileum to the transverse process of L5. The sacroiliac joint is rendered quite sturdy by well developed anterior and posterior sacroiliac ligaments. The pubic symphysis is stabilized by the superior pubic ligament and the inferiorly placed arcuate ligament.
2.	Accessory ligaments: The sacrotuberous and sacrospinous ligaments are important reference structures in pelvic anatomy. The sacrotuberous ligament runs from the sacrum to the tuberosity of the ischium and the sacrospinous ligament runs from the sacrum to the ischial spine. These ligaments form boundaries for the greater and lesser sacral foramina.

B.	Pelvic brim: The bony circle which divides the pelvis into a great portion superiorly (pelvis major) and a smaller portion posteroinferiorly (pelvis minor). Parts of this bony circle include the promontory and ala of the sacrum posteriorly, the arcuate line of the ileum and pectin pubis (iliopectineal line) laterally, and the pubic crest anteriorly.

C.	Major pelvis: The "false" pelvis superior to the pelvic brim.

D.	Minor pelvis: The "true" pelvis inferior to the pelvic brim. Contains pelvic viscera such as urinary bladder, non-pregnant uterus, prostate gland, seminal vesicles etc.

E.	Pelvic wall muscles: These include the obturator internus laterally, whose tendon exits the pelvis via the lesser sacral foramen and the piriformis, posteriorly, whose tendon exits the pelvis via the greater sciatic foramen.

F.	Pelvic diaphragm: The hammock-shaped muscular closure of the pelvis inferiorly. This structure allows passage of the urethra, vagina (female) and anal canal. Anteriorly, the urethra and vagina (if present) exit the pelvis via a gap in the muscular diaphragm, the genital hiatus. Posteriorly, the anal canal is intimately related to the pelvic diaphragm it perforates. The diaphragm consists of three muscles: pubococcygeus and iliococcygeus constitute the major portion of the diaphragm and are referred to as the levator ani. The third muscle in the diaphragm is the relatively small (ischio)coccygeus. The fibers of pubococcygeus that make intimate contact with the posterior aspect of the anorectal junction are referred to as the puborectalis and fibers of the pubococcygeus that sweep around the posterior wall of the vagina and end in the perineal body are called pubovaginalis. Like all muscles, the pelvic diaphragm is covered by fascia. The superior fascia of the pelvic diaphragm is an extension of transversalis fascia; the inferior fascia is an extension of the fascia of obturator internus. The pelvic diaphragm is innervated by branches of $S_{3,4}$.

II. Urinary bladder. Retroperitoneal, capacity approximately 500 mls. The wall of the bladder is composed of smooth muscle fibers, the detrusor muscle. The sensation of the need to void is mediated by GVA fibers accompanying parasympathetic efferent fibers which mediate contraction of the detrusor muscle.

  A. Trigone: The smooth triangular area on the posterior internal aspect of the bladder. Points of triangle=two openings of ureters and urethra. Between the ureters is a fold of mucosa, the interureteric fold.

  B. Urethra: At the bladder neck, there is an internal urethral sphincter which is poorly defined anatomically, but functionally important in males since it prevents retrograde ejaculation. In the male, the urethra is relatively long and has 3 portions: prostatic, membranous, and spongy. The prostatic portion passes through the prostate. The posterior wall in this portion has a longitudinal ridge, the urethral crest. On either side of the urethral crest there are grooves into which prostatic secretions empty. The urethral crest bears a rounded eminence, the seminal colliculus. The prostatic utricle, a blind pouch which is believed to be the homolog of the female uterus and vagina, sits in the middle of the seminal colliculus. Ejaculatory ducts open into the seminal colliculus, on either side of the prostatic utricle. The membranous urethra passes through the urogenital diaphragm. This portion receives the duct of each bulbourethral gland. The central portion of the urogenital diaphragm functions as a voluntary sphincter for the urethra, and is considered a separate muscle, the external urethral sphincter, innervated by the pudendal nerve. The spongy portion passes through the corpus spongiosum of the penis. In the female, the urethra is relatively short (no equivalent of male spongy portion present).

III. Male pelvic viscera.

  A. Prostate: Fibromuscular gland inferior to the urinary bladder, surrounding the first portion of the urethra, like a donut. Two lateral lobes, connected anteriorly by an isthmus. A posterior median lobe, often hypertrophies and blocks urinary flow.

  B. Ductus deferens: Leaves the inguinal canal via the deep inguinal ring, enters the lateral pelvic cavity, and proceeds medially, anterior to the ureter. Near where it crosses the ureter, the ductus deferens enlarges forming the elongated ampulla, and descends medial to the seminal vesicle. Inferiorly, the ampulla abruptly decreases in size, receives the duct of the seminal vesicle, and enters the prostate gland as the ejaculatory duct.

  C. Seminal vesicle: Two finger-like glands, nearly vertical in position, on the posterior aspect of the urinary bladder, lateral to the ampulla of the vas deferens. Their ducts merge with the vas deferens to form the ejaculatory duct, which enters the prostate and opens on the seminal colliculus, near the prostatic utricle.

IV. Female pelvic viscera.

    A.   Vagina: Extends from vestibule to cervix uteri. Space between cervix and vagina= fornix. There are two lateral fornices and an anterior and posterior fornix. The posterior fornix is deeper than the other fornices, and abuts the peritoneum in the pelvis which dips inferiorly forming the rectouterine pouch. The angle between uterus and vagina is approximately 90^0.

    B.   Uterus: Consists of fundus, superior to entrance of oviducts; body, the upper two-thirds of the organ; cervix, lower one third. Supported by peritoneum, forming the broad ligament; the round ligaments, derived from the gubernaculum on each side; and condensations of pelvic fascia forming the important lateral cervical (cardinal) ligaments, the utero-sacral ligaments posteriorly, and the pubocervical ligaments, anteriorly.

    C.   Uterine tubes (Fallopian tubes): Portions include the infundibulum, ampulla, isthmus, and uterine (intramural). Supported by portion of broad ligament termed mesosalpinx.

    D.   Ovary: Sits in a depression of parietal peritoneum on the lateral pelvic wall, termed the ovarian fossa. Attached to uterus via ovarian ligament, derived from gubernaculum. Additional support by a portion of the broad ligament, the mesovarium. Ovarian vessels reach the organ via the suspensory (infundibulopelvic) ligament.

V. Peritoneal reflections in the pelvis.

    A.   In the female, the parietal peritoneum proceeds inferiorly on the deep surface of the anterior abdominal wall, sweeps posteriorly over the dome of the urinary bladder, and then dips inferiorly between the urinary bladder and the uterus, forming the vesicouterine pouch. The peritoneum continues over the dome of the uterus and peritoneum over the inferior and superior aspects of the uterus (two leaves of peritoneum) is pulled laterally on each side, forming the broad ligament. The posterior leaf of broad ligament gives off the mesovarium, which is continuous with the surface epithelium of the ovary. The mesosalpinx is the portion of broad ligament which supports the uterine tubes and the mesometrium is the part of the broad ligament adjacent to the uterus. As peritoneum leaves the uterus posteriorly and inferiorly, it dips deeply inferiorly forming the rectouterine pouch. Peritoneum ascends the posterior aspect of the rectouterine pouch, covering the anterior surface of the rectum in its inferior two-thirds. In the superior third of the rectum, the peritoneum wraps around the structure, forming a mesentery. Thus the superior third of the rectum is mesenteric.

    B.   In the male, the peritoneum follows a course similar to the female, except the uterus is not present. Therefore, the peritoneum leaves the urinary bladder, and has a slight dip inferiorly, forming the rectovesical pouch, before continuing superiorly, covering the rectum.

VI. Vasculature of pelvis.

    A.   The arterial supply to the pelvis is the internal iliac (hypogastric). In addition to pelvic viscera, this artery supplies gluteal branches to the buttocks; the obturator, to the adductors of the thigh; and branches to the perineum. This is one of the most variable arterial systems in the body. A general description of the internal iliac follows: Usually, the artery divides within the pelvis into a small posterior division and a relatively larger anterior division.

        1.   Posterior division: Branches of this division usually include the superior gluteal, which exits the pelvis via the greater sciatic foramen on the superior border of piriformis and supplies the gluteus medius, minimus and tensor fascia lata; the iliolumbar, which ascends and gives branches to the iliacus, psoas major, and quadratus lumborum; and the lateral sacral, which sends branches through the anterior sacral foramina.

2. Anterior division: This major branch provides the inferior gluteal, which exits the pelvis via the greater sciatic foramen on the inferior side of piriformis and supplies the gluteus maximus; the obturator, to medial thigh muscles; the internal pudendal, to perineal structures; and all visceral branches to pelvic viscera. Visceral branches include the umbilical, which gives off the superior vesical proximally, the inferior vesical and middle rectal, which may arise from a common stem, and uterine and vaginal arteries in the female. The uterine artery runs anterior to the ureter (almost a right angle relationship) in the lateral cervical ligament (mnemonic: water runs under the bridge). The vaginal artery may arise directly from the internal iliac or from the uterine artery.

B. Prostatic venous plexus: The plexus of veins in the adventitia of the prostate, which communicates with the vesical plexus of veins. The prostatic plexus receives the deep dorsal vein of the penis. The prostatic plexus drains into the internal iliac vein. There may be connections with the vertebral venous plexus and this may be a route by which cancer can spread from the prostate to the vertebral column.

C. Lymphatics: In general, lymph from pelvic viscera tends to drain towards internal iliac and sacral nodes. Lymphatics from the perineum, external genital structures, lower two thirds of vagina, and anal canal inferior to the pectinate line of the anal canal, drain into superficial inguinal nodes. Lymphatics from upper one third of the vagina drain into internal iliac nodes. Some lymphatics from the body of the uterus run with the round ligament of the uterus through the inguinal canal to reach superficial inguinal lymph nodes. Lymphatics from the ovaries and testis drain into para-aortic nodes.

VII. Autonomic nervous system in pelvis.

A. Parasympathetic: The parasympathetic supply to the GI tract inferior to the left colic flexure and pelvic viscera is via the pelvic splanchnic nerves from S_{2-4}. The gonads may receive some parasympathetic contributions from the vagus.

B. Sympathetic trunks and ganglion impar: Abdominal sympathetic trunks continue caudally into the pelvis by passing over the ala of the sacrum, medial to anterior sacral foramina and converging into the ganglion impar near S5. Two or three sacral splanchnic nerves (sympathetic) are given off which enter the inferior hypogastric plexus.

C. Superior hypogastric plexus, hypogastric nerves, and inferior hypogastric plexus: This is the caudal continuation of the prevertebral (preaortic) plexus inferior to the bifurcation of the aorta into common iliac arteries. This plexus receives an "injection" of sympathetic GVE components via the two lower lumbar splanchnic nerves. At the sacral promontory, the superior hypogastric plexus bifurcates into right and left hypogastric nerves which continue into the pelvis and join the inferior hypogastric (pelvic) plexus on each side of the rectum. The inferior hypogastric plexus also receives parasympathetic contributions from pelvic splanchnic nerves and GVA components are also present. Mixed branches from the inferior hypogastric plexus distribute with branches of the internal iliac artery to pelvic viscera.

VIII. Pudendal nerve: Somatic nerve formed from anterior divisions of S_{2-4} and is the principal nerve of the perineum. The pudendal nerve has the four components of spinal nerves, viz., GSE, GVE, GSA and GVA. Parasympathetic (and perhaps some sympathetic) GVE fibers pass to the perineum from the prostatic (male) or uterovaginal plexus (female). The pudendal nerve leaves the pelvis via the greater sciatic foramen, goes dorsal to the ischial spine, re-enters the pelvis via the lesser sciatic foramen and then enters the pudendal canal on the lateral wall of the ischioanal fossa. The pudendal nerve divides into 3 terminal branches in the pudendal canal. The first of these branches, the inferior rectal, supplies the external sphincter ani and adjacent skin and anal canal inferior to the pectinate line. Another branch, the perineal, divides into posterior scrotal (labial) and deep perineal nerves. The posterior scrotal (labial) is sensory and the deep perineal innervates skeletal muscles in the UG triangle. The remaining branch of the pudendal, the dorsal nerve of the penis (clitoris), is purely sensory from the glans, prepuce and skin of the penis (clitoris).

IX. Functional anatomy.

   A.   Erection is mediated by parasympathetic nerves innervating the penis which cause engorgement by dilation of arteries and/or relaxation of their usual vascular tone. A male whose spinal cord has been severed superior to sacral levels may still be capable of erection but he may not "feel" stimulation of the penis. The female equivalent of erection is vaginal lubricity, the vaginal "sweating" resulting from vascular engorgement of the vaginal bulb subsequent to sexual arousal.

   B.   Ejaculation is mediated by the sympathetic system. Sympathetics cause closure of the functional sphincter at the neck of the bladder, and mediate contractions of smooth muscle in the male duct system to move semen into the membranous urethra. This process is referred to as emission. External release of semen (ejaculation proper) is accompanied by clonic spasms of the bulbospongiosus and ischiocavernosus muscles. The female equivalent of ejaculation is orgasm which involves rhythmic contractions of the orgasmic platform, and contractions from the uterine fundus inferiorly. These effects are also believed to be mediated by the sympathetic nervous system.

X. Perineum. The diamond-shaped region around the outlets of the genital, urinary, and gastrointestinal tract. For descriptive purposes, this diamond-shaped area is divided into a posterior anal triangle and an anterior urogenital triangle.

   A.   Anal triangle: Important components include the anus, the opening of the anal canal; the external anal sphincter, the large voluntary sphincter surrounding the inferior two-thirds of the anal canal; and ischioanal (ischiorectal) fossa.
      1.   Anal canal: The terminal 2 inches of the GI tract, inferior to the puborectal sling. The anal canal is oriented at a right angle to the rectum. The superior portion of the canal has longitudinal bulges of mucosa called anal columns. Inferiorly, the anal columns are united by mucosal folds called anal valves. Inferior to anal valves there is a zone of hairless skin called the pectinate line which marks the site of the breakdown of fetal hindgut (endoderm) and proctodeum (ectoderm) to form the anal opening. The pectinate line is the line of demarcation between visceral and somatic supply to the anal canal. Superior to the pectinate line, lymphatics drain cranially into internal iliac nodes, sensory and motor nerve supply is visceral, and blood supply is from superior rectal branch of inferior mesenteric venous return is via hepatic-portal system. Inferior to the pectinate line, the nerve supply is somatic, lymphatics drain into inguinal nodes, and the arterial supply is from inferior rectal (pudendal) or middle rectal (internal iliac) arteries. Venous return is via caval system.
      2.   Ischioanal (ischiorectal) fossa: The area on either side of the anal canal containing fat, connective tissue, and vessels. The fat in this fossa is primarily liquid at body temperatures and allows the anal canal to dilate at defecation. The lateral wall of this fossa is the obturator internus and ischial tuberosity; the superomedial wall is levator ani and external sphincter ani. Internal pudendal vessels and nerve travel on the lateral

wall in a duplication of the fascia covering obturator internus, the pudendal (Alcock's) canal. Inferior rectal branches of the pudendal vessels and nerve traverse the fat and connective tissue in the ischioanal fossa to supply the external sphincter ani.

B.   Urogenital triangle. The anterior triangle of the perineum, contains the outlets of the urinary and genital systems.

    1.   Urogenital diaphragm: The muscular diaphragm, inferior to the genital hiatus of the pelvic diaphragm, that covers the urogenital triangle. The muscle in the diaphragm is the deep transverse perineus. The part of deep transverse perineus that surrounds the urethra passing through is the voluntary external urethral sphincter, innervated by the pudendal nerve.

    2.   Superior fascia of urogenital diaphragm: By some accounts, this fascia is formed as an extension of transversalis fascia, sweeping inferiorly through the genital hiatus of the pelvic diaphragm, and reflecting laterally as the superior fascia of the urogenital diaphragm. Others consider this fascia as incomplete, and best developed only around the external urethral sphincter.

    3.   Inferior fascia of urogenital diaphragm: Also called the perineal membrane. By all accounts, a well developed fascial sheet, attaching laterally to the ischiopubic rami, and posteriorly along the base of the urogenital diaphragm.

    4.   Deep perineal pouch (space): The potential space superior to the perineal membrane. In both sexes, the external urethral sphincter and membranous urethra are located here. Additionally, in the male, the bulbourethral gland is in this space.

    5.   Superficial perineal pouch (space): Recall that Colles' fascia, the continuation of Scarpa's fascia into the perineum, attaches laterally to the ischiopubic rami and posteriorly to the base of the urogenital diaphragm. The potential space between Colles' fascia and the perineal membrane is the superficial pouch. It houses the root (immobile attached portion) of the penis (clitoris) and associated muscles, and in the female, the bulb of the vagina and the greater vestibular gland. Recall that the homolog of the greater vestibular gland in the male, the bulbourethral (Cowper's) gland, resides in the deep perineal space. In both sexes, the fragile superficial transverse perineus, resides in the superficial space.

        a)   Contents in male: erectile tissue attachments and associated muscles: The proximal portion of the corpus cavernosum, termed a crus, is on either side attached to the inferior pubic ramus and covered by the ischiocavernosus muscle. Medially, the bulb of corpus spongiosum, attached to the perineal membrane, and covered by the bulbospongiosus. Lastly, the superficial transverse perineus muscle, originating from the ischial tuberosity and inserting medially into the perineal body. All these muscles are innervated by branches of the pudendal nerve.

        b)   Contents in female: Vestibular bulbs are erectile bodies on either side of the vestibule of the vagina covered by bulbospongiosus muscle. Greater vestibular glands reside at the posterior aspect of the bulb of the vagina and their ducts open into the vestibule. The perineal body also occurs in males, but it has a more significant supporting role in females. Also called the central tendinous point of the perineum, it is the connective tissue structure between the anus and posterior vaginal wall formed by interdigitation of attaching fibers of the bulbospongiosus, superficial transverse perineus, external anal sphincter, deep transverse perineus and levator ani.

C. Male genitalia.
   1. Penis: The penis has 3 columns of erectile tissue: There are two columns of corpus cavernosum attached laterally in the superficial space to the inferior pubic ramus and the medially placed single column of corpus spongiosum, which transmits the urethra. Proximally the corpus spongiosum is enlarged forming the bulb, distally it is enlarged forming the glans of the penis. Tunica albuginea binds the 3 columns of erectile tissue. Superficial to tunica albuginea is the deep fascia of the penis. The suspensory ligament, attaches to the penis at the juncture of the root and shaft (mobile portion). This ligament is attached to the underside of the pubic arch and pubic symphysis.
   2. Scrotum: The cutaneous sac which houses the testis.
   3. Testis: Sits in the scrotal sac. Its anterior border is free, the posterior border has the epididymis attached. The tail of the epididymis continues as the vas deferens. The dense connective tissue capsule of the testis is the tunica albuginea. Adherent, to, the anterior half of the testis is a remnant of the processes vaginalis, the tunica vaginalis. The tunica vaginalis extends from superior to inferior poles of the organ.

D. Female genitalia.
   1. Clitoris: Similar to the penis in structure, but much smaller and does not transmit the female urethra.
   2. Labia majora: Female homologue of the scrotum.
   3. Labia minora: Female homologue of skin and connective tissue on ventral surface of penis. Splits anteriorly into superior and inferior folds which form the prepuce and frenulum of the clitoris, respectively.
   4. Vestibule: The space between the labia minora. Opening into the vestibule are the vagina, urethra, ducts of paraurethral and greater vestibular glands.

XI. Applied anatomy/anatomical pearls.

   A. External hemorrhoids: Varicose enlargements of inferior rectal vein inferior to pectinate line. Painful because of somatic innervation.

   B. Internal hemorrhoids: Varicose enlargements of middle or superior rectal veins superior to pectinate line. Less painful because visceral sensory innervation is less sensitive than somatic sensory innervation.

   C. Pudendal block: To block the pudendal nerve for obstetrical purposes, the needle is inserted through the vaginal wall or just lateral to the labia majora at the level of the ischial tuberosity. The nerve is blocked at the ischial spine, prior to its entrance into the pudendal canal.

   D. Posterior fornix of vagina: A needle is easily passed through this region to obtain fluid samples from the rectouterine pouch for analysis.

   E. Lymphatic drainage: From the testis to para-aortic nodes in the abdominal cavity. From the scrotum and penis to inguinal nodes.

   F. Pectinate line in anal canal: Like a watershed area; superior to this line, lymphatics drain cranially to internal iliac nodes, vessels are from superiorly, and innervation is visceral. Inferior to this line, lymphatics drain to inguinal nodes, vessels are from inferior rectal branches of pudendal, and innervation is somatic.

   G. Ureter and uterine artery at base of broad ligament: Remember, water (ureter) runs under the bridge (of the uterine artery).

   H. Female urethra: Short length predisposes to infections of urinary tract.

I.   Female vs male peritoneal cavity: Male is closed. Female opens to outside via ostia of uterine tubes.

LOWER LIMB

I.   Embryological considerations.

   A.   Limb rotation: After formation of limb buds, the axis of the limb rotates 90^0 medially so the preaxial (flexor) musculature comes to lie posteriorly and postaxial (extensor) musculature moves anteriorly.

   B.   Innervation pattern: In the lower limb, preaxial muscles are innervated by nerves formed from anterior divisions of ventral primary rami, postaxial muscles are innervated by nerves formed from posterior divisions of nerves in the lumbo-sacral plexus. Relative to the thigh and gluteal region, it is also helpful to remember that muscles originating from the pubis and ischium are preaxial and innervated by anterior division (preaxial) nerves, and muscles originating from the ileum and femur are postaxial and innervated by posterior division (postaxial) nerves.

II.   Lumbo-sacral plexus: The key to understanding functional anatomy of the lower limb is understanding the major elements of the nerve supply. The somatic supply to the lower limb is derived from the lumbo-sacral plexus. This plexus, like the brachial plexus, allows for nerves from different segmental levels to merge and form peripheral nerves containing components from several levels. The lumbo-sacral plexus is formed from ventral primary rami from lumbar and sacral levels. These ventral primary rami divide into anterior and posterior divisions. Subsequently anterior division components from different levels converge to form anterior division nerves which innervate preaxial muscles and posterior division components converge to form posterior division nerves which innervate postaxial muscles.

   A.   Anterior division nerves:
      1.   Iliohypogastric and ilioinguinal from L_1. Recall that the iliohypogastric is sensory to the suprapubic region and the ilioinguinal exits the superficial inguinal ring to provide sensation to the scrotum and medial thigh.
      2.   Genitofemoral ($L_{1,2}$). Motor to cremaster, sensory to medial thigh.
      3.   Obturator ($L_{2,3,4}$). Motor to medial compartment of thigh, sensory to medial thigh.
      4.   Accessory obturator ($L_{2,3}$). When present, innervates the pectineus.
      5.   Tibial ($L_{4,5}S_{1,2,3}$). The preaxial component of the sciatic in the thigh, innervates preaxial muscles of thigh and leg.
      6.   Nerve to quadratus femoris and inferior gemellus ($L_{4,5}S_1$).
      7.   Nerve to obturator internus and superior gemellus ($L_5S_{1,2}$).
      8.   Posterior femoral cutaneous (anterior $S_{2,3}$; posterior $S_{1,2}$). This large cutaneous nerve of the posterior thigh has both anterior and posterior division components.
      9.   Pudendal ($S_{2,3,4}$). The great somatic nerve to the anal and urogenital triangles.
      10.   1Nerve to levator ani and coccygeus ($S_{3,4}$).

   B.   Posterior division nerves:
      1.   Lateral femoral cutaneous ($L_{2,3}$). Cutaneous to lateral thigh.
      2.   Femoral ($L_{2,3,4}$). The great motor nerve to the anterior compartment of the thigh, also provides sensory branches to anterior thigh.
      3.   Superior gluteal ($L_{4,5}S_1$). Motor to gluteus medius and minimus, tensor fascia lata.
      4.   Inferior gluteal ($L_5S_{1,2}$). Motor to gluteus maximus, the powerful extensor of the thigh.

5. Common peroneal ($L_{4,5}, S_{1,2}$). The postaxial component of the sciatic in the thigh. Motor to anterior and lateral compartments of leg.
6. Nerve to piriformis ($S_{1,2}$).
7. Perforating cutaneous nerve ($S_{2,3}$). Pierces sacrotuberous ligament, sensory.
8. Perineal branch of S_4. Sensory to skin over ischioanal fossa.

III. Dermatomes: Dermatomes of the lower extremity are important, particularly for diagnostic purposes. Review a diagram of dermatomes in one of the standard atlases.

IV. Overview of blood supply: The gluteal region receives blood supply from the internal iliac artery via the gluteal arteries. The internal iliac also gives off the obturator which contributes to the blood supply to the medial thigh. The major arterial flow to the lower limb is via the external iliac, which changes names to the femoral, inferior to the inguinal ligament. The femoral provides blood supply to all compartments of the thigh, continues to the posterior aspect of the knee as the popliteal, which provides anterior and posterior tibial arteries to supply the leg and foot. The peroneal branch of the posterior tibial nourishes the lateral compartment of the leg.

V. Thigh: The great saphenous vein is located on the medial thigh and passes through the fascia lata, the investing fascia of the thigh, on the cranial, medial aspect of the thigh and drains into the femoral vein. The fascia lata has deep extensions which divide the thigh into 3 compartments: An anterior, medial, and posterior compartment.

A. Medial compartment: This preaxial adductor compartment is, innervated by the obturator nerve. Muscles in this compartment include the gracilis, most superficial on the medial aspect of the thigh; pectineus and adductor longus, also superficial, but more anterolateral in position; adductor brevis, deep to adductor longus; adductor magnus, deepest member of this compartment. The cranial portion of adductor magnus, with horizontally oriented fibers, takes origin from the ischial tuberosity, therefore, this portion of adductor magnus is innervated by the "hamstring nerve", the tibial.

B. Anterior compartment: This postaxial compartment includes the sartorius and quadriceps femoris (rectus femoris, vastus lateralis, medialis, and intermedius), both innervated by the femoral nerve. The quadriceps is the powerful extensor of the leg and its rectus femoris portion also flexes at the hip. The long sartorius flexes the knee and thigh, and laterally rotates the thigh. The powerful flexor of the thigh, the iliopsoas, enters this compartment cranially on its way to insert on the lesser trochanter.

C. Posterior compartment: This preaxial compartment contains the hamstring muscles; the biceps femoris, semimembranosus and semitendinosus. In general these muscles extend the thigh and flex the knee. They are innervated by the tibial portion of the sciatic nerve, except the short head of the biceps femoris, (postaxial) which originates from the femur and is innervated by the common peroneal portion of the sciatic nerve.

D. Femoral triangle: An important reference area on the anteromedial thigh.
   1. Boundaries: The base of the triangle, superior in position, is the inguinal ligament, medial side is adductor longus and lateral side is sartorius. The floor of the triangle, from lateral to medial, is the iliopsoas, pectineus, and adductor longus muscles.
   2. Contents: Entering the base of the triangle by passing deep to the inguinal ligament are from lateral to medial: femoral nerve (lying on iliacus), femoral artery (lying on psoas major), femoral vein (near junction of psoas and pectineus), and femoral canal for lymphatic vessels and several lymph nodes. Mnemonic= NAVL. Important branches of

the femoral vessels in the triangle include: medial and lateral femoral circumflex and profunda femoris.

3.   Femoral hernia: protrusion of abdominal viscus through the femoral canal, displacing lymphatics. Always deep to inguinal ligament. Relatively uncommon hernia, but potentially dangerous because contents may become strangulated by taut inguinal ligament.

4.   Adductor (subsartorial) canal (of Hunter): Begins at the apex of the femoral triangle and ends at the adductor hiatus, the opening in the tendon of adductor magnus leading to the popliteal fossa. This canal contains the femoral artery and vein, and two branches of the femoral nerve: the nerve to vastus medialis and the saphenous nerve.

VI. Gluteal region.

A.   Gluteus maximus and tensor fascia lata: Gluteus maximus=the large, rhomboid shaped, posterior, superficial, powerful extensor of the thigh, innervated by inferior gluteal nerve. Tensor fascia lata=laterally placed between two leaves of fascia lata, originates superiorly from lateral iliac crest and caudally inserts into iliotibial tract. An important abductor of thigh.

B.   Deep gluteal musculature: Under cover of the vast gluteus maximus, is the gluteus medius, which covers the deeply placed gluteus minimus. At the inferior edge of the gluteus medius is the piriformis, an important reference muscle for the anatomy of this region. (Superior gluteal nerve and vessels exit the pelvis superior to the muscle, inferior gluteal, sciatic, and posterior femoral cutaneous nerves and inferior gluteal vessels exit the pelvis inferior to this muscle.) Inferior to the piriformis, is the obturator internus with its two gemelli and most inferiorly is the quadratus femoris. The tendon of obturator externus, one of the medial thigh compartment muscles, can be seen deep, between inferior gemellus and quadratus femoris.

C.   Gluteus medius and minimus and walking: These muscles hold the pelvis horizontal when support is removed on the contralateral side, such as when walking or standing on one leg. If the pelvis dips inferiorly on the side not supported, there is a problem with the gluteus medius and minimus on the opposite side. In such situations, this deficit is referred to as a positive Trendelenburg sign.

D.   Sciatic nerve: The sciatic nerve leaves the pelvis via the greater sciatic foramen on the inferior border of piriformis and descends through the gluteal region midway between the ischial tuberosity and greater trochanter. To avoid hitting the sciatic nerve during intramuscular injections, the needle is introduced into the superior, lateral quadrant of gluteus maximus.

E.   Hip joint: The large femoral head articulates in the well developed socket, the acetabulum. Well developed ligaments, reinforcing the capsule, limit movement at this joint and consideration of the arterial supply to the joint is important because of the frequency of fractures of the femoral neck, particularly in the elderly.

1.   Ligaments: A well developed, anteriorly placed, Y-shaped iliofemoral ligament prevents hyperextension; the posteriorly placed ischiofemoral ligament also helps prevent hyperextension; and the pubofemoral ligament helps limit abduction and extension. Within the joint, the ligament of the head may help limit adduction.

2.   Arterial supply: From medial and lateral circumflex, and inferior extensions of superior and inferior gluteal arteries. The obturator artery gives off a branch that accompanies the ligament of the head to the femoral head. This branch is well developed in children

but in the elderly, following fractures of the proximal femoral neck, this branch may not be sufficient to maintain the femoral head, and aseptic necrosis of the head may follow.

VII. Knee.

A. Bones: Three bones articulate at the knee: the condyles of the femur and tibia and the patella. The patella articulates only with the femur. The patella develops as a sesamoid bone in the tendon of quadriceps femoris.

B. Menisci: A medial and lateral fibrocartilaginous meniscus (semilunar cartilage) is firmly attached to the medial and lateral flattened condyles (plateaus) of the tibia.

C. Major ligaments.
   1. Collateral: A tibial and fibular collateral ligament which provide side to side stability to the knee joint. The tibial collateral ligament is firmly attached to the medial meniscus. Thus, when this ligament is stretched, it usually affects the attached meniscus. The fibular collateral ligament is separated from the lateral meniscus by the tendon of the popliteus muscle.
   2. Cruciate: Anterior and posterior cruciate ligaments, crossed and named by their attachment to the tibia, are deep to the capsule of the knee joint. They firmly bind the femur and tibia and help limit anterior and posterior displacement of the two bones. The integrity of the ligaments is tested as follows: The patient is seated and the knee is flexed 90°. The anterior cruciate is tested by attempting to pull the tibia anteriorly on the femoral condyles. Similarly, the posterior cruciate is tested by attempting to push the tibia posteriorly on the femoral condyles. The movements in this examination are referred to as "drawer movements."

D. Bursae: The large bursa deep to the quadriceps tendon, the suprapatellar bursa, usually connects with the synovial cavity of the joint. The prepatellar bursa is usually well defined and inferiorly, deep to the quadriceps tendon, there is a smaller deep infrapatellar bursa.

E. Popliteal fossa: The posterior aspect of the knee. Contains popliteal vessels, small saphenous vein, tibial and common peroneal nerves, and lymph nodes. The floor of the fossa is the popliteus muscle.

VIII. Leg.

A. Superficial features. Cutaneous structures include the great saphenous vein medially, accompanied by the saphenous nerve; and the lesser saphenous vein, emerging posterior to the lateral malleolus, and ascending the posterolateral aspect of the leg to drain into the popliteal vein. The lesser saphenous vein is accompanied by the sural nerve in the leg. The inferior continuation of the fascia lata into the leg is the crural fascia. This fascia divides the leg into anterior, lateral, and posterior compartments, each with its own blood and nerve supply.

B. Anterior compartment: This postaxial compartment contains three muscles: the tibialis anterior, extensor digitorum hallucis, and extensor digitorum. A divergent lateral slip of extensor digitorum inserting on the base of the fifth metatarsal, the peroneus tertius, is classified as a fourth muscle by some authors. The tibialis anterior dorsiflexes and inverts the foot; the names of the other muscles indicate their function. The artery to the anterior compartment is the anterior tibial branch of popliteal. The anterior tibial enters the compartment superior to the interosseus membrane. The deep peroneal branch of the common peroneal nerve provides innervation to this compartment.

C. Lateral compartment: This postaxial compartment contains two muscles which evert the foot: The peroneus longus and brevis. Tendons of both muscles pass posterior to the lateral malleolus. The peroneus longus lies superficial to and covers the brevis. The artery to this compartment, the peroneal, is a branch of the posterior tibial. The nerve to the compartment is the superficial peroneal. After innervating the two muscles in the compartment, the nerve becomes cutaneous in the lower one-third of the leg and continues onto the dorsum of the foot.

D. Posterior compartment: This preaxial compartment has superficial and deep layers. All muscles in this compartment are innervated by the tibial nerve, and blood supply to this compartment is from the posterior tibial branch of the popliteal.

1. Superficial posterior: All three of the muscles here (gastrocnemius, soleus, and plantaris) insert into the calcaneal tendon and are powerful plantar flexors.

2. Deep posterior: Muscles include from lateral to medial, flexor hallucis longus, tibialis posterior, and flexor digitorum longus. Tendons of these muscles enter the medial side of the plantar surface of the foot. The function of the muscles is indicated by their names, except for tibialis posterior. This muscle plantar flexes and inverts the foot.

IX. Ankle.

A. Retinacula: Connective tissue structures which maintain the mechanical advantage of muscle tendons by holding the tendons against bone.

1. Extensor: A superior extensor retinaculum, extends from tibia to fibula just cranial to the medial and lateral malleolus. An inferior extensor retinaculum covers the dorsum of the foot. This inferior extensor retinaculum is Y shaped and the leg of the Y attaches laterally to the calcaneus. Medially, one diverging arm attaches to the medial malleolus and the other arm goes inferiorly and attaches to plantar fascia. The inferior edge of the inferior extensor retinaculum is the site where the anterior tibial artery changes names to dorsal pedis. Dorsalis pedis is an important artery for taking pulse in the foot.

2. Peroneal: There is a superior and inferior retinaculum holding the two lateral compartment tendons down on the calcaneus as they pass posterior to the lateral malleolus.

3. Flexor: Extends from the medial malleolus to the calcaneus and holds down the tendons of the flexors of the deep posterior compartment and tibial nerve and posterior tibial vessels.

B. Medial malleolus and associated structures. This includes the structures deep to the flexor retinaculum. Tendons of tibialis posterior and flexor digitorum longus pass posterior to the medial malleolus but flexor hallucis longus, from the lateral leg, arches more inferiorly and uses the sustentaculum tali of the calcaneus as a pulley.

X. Dorsum of foot.

A. Dorsalis pedis artery: the continuation of the anterior tibial artery, inferior to the extensor retinaculum, across the dorsal arch of the foot.

B. Extensor digitorum brevis: Arises from calcaneus and assists the extensor digitorum. Its most medial slip is termed the extensor hallucis brevis. These muscles are innervated by the deep peroneal nerve. This nerve continues distally to provide sensation to the dorsum of the great and second toes and the webbed area proximally between these two toes.

XI. Plantar foot.

    A.   Cutaneous nerves. Cutaneous branches around the ankle and plantar foot include the medial calcaneal, and medial and lateral plantar nerves, all three which are branches of the tibial nerve.

    B.   Lamination. The plantar surface of the foot is usually described in layers.

        1.   Superficial layer: This layer is deep to the plantar aponeurosis. From lateral to medial, muscles in this layer are the abductor digiti minimi, innervated by lateral plantar nerve; and flexor digitorum brevis and abductor hallucis, innervated by medial plantar nerve. The flexor digitorum brevis is homologous to the flexor digitorum superficialis in the upper limb.

        2.   Second layer: Muscles in this layer are flexor hallucis longus, flexor digitorum longus, the four lumbricals and quadratus plantae. The first lumbrical is innervated by medial plantar nerve, the lateral three lumbricals and quadratus plantae by lateral plantar nerve. The flexor hallucis longus and flexor digitorum are innervated in the leg by the tibial nerve.

        3.   Third layer: Muscles in this layer are flexor digiti minimi and adductor hallucis, innervated by lateral plantar nerve and the flexor hallucis brevis, innervated by the medial plantar nerve.

        4.   Fourth layer: Includes the three plantar and four dorsal interossei anteriorly, and the tendons of peroneus longus and tibialis posterior, posteriorly. All interossei are innervated by the lateral plantar nerve.

    C.   Summary of motor innervation of intrinsic muscles of plantar foot: The lateral plantar nerve mirrors the ulnar in the hand and the medial plantar nerve is similar to the median. Therefore, the medial plantar nerve innervates the flexor hallucis brevis, abductor hallucis, first lumbrical and flexor digitorum brevis. All other intrinsic muscles of the foot are innervated by lateral plantar nerve.

    D.   Ligaments: There are numerous ligaments in the foot, named primarily according to location. Important ligaments include the deltoid, the medial ligament of the ankle, which attaches the medial malleolus to the tarsus; the lateral ligament of the ankle, which includes the calcaneofibular and anterior and posterior talofibular ligaments; and on the plantar surface, the spring ligament or calcaneonavicular, and the long plantar ligament. The plantar ligaments help maintain the arch of the foot.

XII. Applied anatomy.

    A.   Fractures of pelvis: Since the pelvis is a ring-like structure, fractures typically occur in more than one location. Example: through both the pubic symphysis and sacroiliac.

    B.   Gluteal injections: To avoid injury to the sciatic nerve, injections should be made into the superior, lateral quadrant of the buttock.

    C.   Common peroneal nerve: Vulnerable to injury (falls, tight leg cast) as it winds around neck of fibula. Injury causes foot drop.

    D.   Knee injury: A blow to the posterolateral side of the knee may result in a torn tibial collateral ligament, medial meniscus and anterior cruciate ligament (the unhappy triad).

    E.   Ankle sprains: Most are inversion injuries since the medial malleolus does not extend as far inferiorly as the lateral malleolus.

F.   Dorsalis pedis artery: Easily accessible for determining pulse in the foot.

G.   Trendelenburg sign: To hold the pelvis level when standing on one leg, the abductors (gluteus medius, minimus and tensor fascia lata) of the hip on the side supporting the weight must contract to hold the pelvis horizontal. If these muscles are weakened, the pelvis tilts inferiorly on the opposite side. This tilting is a positive Trendelenburg sign.

NEUROANATOMY

HISTOLOGY OF THE NERVOUS SYSTEM

I. The basic functional unit of the nervous system is the neuron.

    A. Principle parts of a neuron are:

        1. the <u>perikaryon</u>, also termed the cell body or soma. The cell body contains all of the synthetic machinery for maintaining the cell (nucleus, one or more nucleoli, mitochondria, ribosomes, etc.) and for manufacturing and packaging neurotransmitters (rough endoplasmic reticulum, Golgi apparatus, neurotransmitter-containing synaptic vesicles, etc). A unique feature of neurons is the high concentration of aggregates of RNA within the cell body. When stained with a basic dye, such as Thionin or Toluidine Blue, the aggregates of RNA appear "pepper-like" and are termed Nissl substance.

        2. the <u>axon</u> which is responsible for carrying the impulse, or action potential, away from the cell body. Axons contain neurofilaments, axoplasm and neurotubules, which act as highways for movement of synaptic vesicles and other organelles, such as mitochondria. The axon is continuous with the cell body through the axon hillock. This area is unique in that it is devoid of Nissl substance, thus appearing pale in stained sections. Neurons possess only one axon, however axonal branching (collaterals) is common.

        3. the <u>axon terminal</u>, which is the expanded end of the axon, contains axoplasm, mitochondria, synaptic vesicles and specializations along portions of the axolemma (term used to describe the plasma membrane of neurons) known as synapses. Synapses are select areas where vesicles are able to fuse with the axolemma resulting in the discharge of the neurotransmitter into the synaptic cleft, which is the region between the pre- and post-synaptic membranes (approximately 20 nm in width). Released neurotransmitter then binds to receptors on the post-synaptic membrane resulting in the initiation of an action potential in the "down-stream" neuron. Synapses may occur between axons and dendrites (axo-dendritic), neuronal cell bodies (axo-somatic), other axons (axo-axonal) or skeletal muscle cells. Concerning the innervation of smooth and cardiac muscle, axon terminals with synaptic specializations are rare. Instead, vesicles are released at expansions which occur at frequent intervals along the axon, thus resembling a "string of pearls". Morphologically, these expansions are described as boutons-en-passage.

        4. the <u>dendrites</u> which are responsible for carrying impulses toward the cell body. Neurons may have one or more dendrites and each dendrite may branch extensively. Dendrites typically do not contain synaptic vesicles.

    B. Neuronal morphology

        1. Neurons come in a myriad of sizes ranging from the granule cells of cerebellum which are approximately 5 mm in diameter to the giant cells of Betz in area 4 of the cerebral cortex which may be 120 mm in diameter.

        2. Neurons also vary remarkably in shape. Morphologically, neurons can be classified into one of the following categories.

            a) Unipolar neurons are devoid of dendrites and possess only one process, an axon. Unipolar neurons are typical in the immature nervous system.

b) Bipolar neurons possess one dendrite and one axon. Examples include the sensory neurons supplying the vestibular apparatus and the cochlea [cranial nerve (CN) VIII].

c) Pseudounipolar neurons have a single process that branches, thus forming a "T". These neurons originate as bipolar neurons, but during development the two processes migrate towards each other and eventually fuse. Examples of pseudounipolar neurons are sensory neurons within dorsal root and cranial nerve ganglia (except CN VIII). Many authors include pseudounipolar neurons in the unipolar category.

d) Multipolar neurons comprise the largest population of neurons in the adult nervous system and contain two or more dendrites and a single axon. Examples of multipolar neurons include motor neurons in the spinal cord, Purkinje neurons of the cerebellum and pyramidal neurons of the cerebral cortex.

3. Neurons are also classified according to whether or not they possess long axons (Golgi type I) or short axons (Golgi type II).

II. Non-neuronal cells of the nervous system

A. A unique feature of nervous tissue proper is that it does not possess connective tissue elements such as fibroblasts or collagen. Rather, support for neurons is provided by glial (neuroglial) cells. Interestingly, glial cells out number neurons about 10 to 1, but comprise only about 50% of the volume of the nervous system.

B. Glial cells are characterized according to the following location and/or function.

1. **Central Nervous System**

a) Astrocytes - Processes of astrocytes are commonly located adjacent to blood vessels, neurons and beneath the ependyma lining the ventricular system and the pia mater on the outer surfaces of the brain and spinal cord. Their functions include providing mechanical support to the CNS, assisting in providing nutrition to neurons and relaying cellular signals in a non-synaptic manner to neurons. Two types of astrocytes exist. Fibrous astrocytes are characterized by a relatively small cell body, but contain an extensive array of processes. Fibrous astrocytes are located predominantly in the white matter. The cell body of protoplasmic astrocytes is larger than that of fibrous astrocytes. Protoplasmic astrocytes also contain numerous processes and are located predominantly in the grey matter. Following an injury to the CNS, astrocytes become "reactive", invade the injury site and contribute to the formation of a glial scar.

b) Oligodendrocytes - Form myelin in the CNS. Myelin of myelinated axons is formed by oligodendrocytes that wrap axons with concentric lamella of their plasma membrane. The cytoplasm is extruded from the lamella resulting in layer upon layer of plasma membranes. One oligodendrocyte will wrap only a portion of several axons with one axon being wrapped by many oligodendrocytes. One or more unmyelinated axons will still be surrounded by an oligodendrocyte, however, the glial plasma membrane does not form into concentric lamellae.

c) Microglia - These glial cells are unique in that they are the only ones not derived from ectoderm, being derived instead from mesoderm. Their function is to act as scavengers/macrophages following an injury. While it is difficult to locate these small cells in normal CNS tissue, they are abundant following an injury.

d) Ependyma - These cells of ectodermal origin that line the ventricular system of the brain, brainstem and the central canal of the spinal cord. In the embryo, ependymal cells proliferate, thus giving rise to all neurons and glial cells (except microglia). The choroid plexus, which produces cerebrospinal fluid, is also considered ependymal tissue.

2. **Peripheral Nervous System**

   a) Satellite cells - Neuronal cell bodies outside of the brain and spinal cord, i.e., primary afferent neurons in cranial nerves, sensory and autonomic ganglia, are surrounded by satellite cells which are of ectoderm origin. Satellite cells are similar to astocytes and oligodendrocytes of the CNS.

   b) Schwann cells - <u>Myelinated axons</u> in peripheral nerves are surrounded by sheaths of myelin from Schwann cells which are ectodermal in origin. One Schwann cell may ensheathe, or myelinate only one axon. Small pockets of cytoplasm are often seen within the concentric lamellae of the Schwann cells. These pockets, termed clefts of Schmidt-Lanterman, are tunnels of cytoplasm extending from the inner lamella to the outer lamella. The area of an axon whereby the myelin of one Schwann cell ends and another begins is known as a node of Ranvier. As is the case with oligodendrocytes in the CNS, <u>unmyelinated axons</u> of peripheral nerves are still enclosed in a single lamella of Schwann cell plasma membrane.

III. Arrangement of Peripheral Nerves

Each axon within a peripheral nerve is surrounded by a Schwann cell. The Schwann cell in turn, secretes and is surrounded by, a basal lamina and a thin layer of reticular fibers. The basal lamina and the layer of reticular fibers are termed the <u>endoneurium</u>. Bundles of axons with their endoneurial coverings are group together into fascicles. Fascicles within a peripheral nerve are surrounded by concentric layers of connective tissue termed the <u>perineurium</u>. Finally, fascicles are arranged into nerves which have a strong outer covering of connective tissue termed the <u>epineurium</u>. These connective tissues are important in nerve regeneration and surgical reattachments.

IV. Reaction of neurons to injury

A. Ischemic, traumatic or chemical insults to neurons result in a characteristic response known as chromatolysis. Hallmarks of chromatolytic neurons include:

   1. a nucleus that is eccentric in position within the cell body. Under normal conditions, the nucleus is usually centrally located.

   2. dispersion of the cytoplasmic aggregates of RNA. This results in a pale appearance when using Nissl stains.

   3. swelling of the neuronal cell body.

B. When a myelinated axon is severed, the proximal stump degenerates in a retrograde direction a distance of one or two node of Ranvier segments. The distal axonal segment, having been separated from the cell body, degenerates. This processes is known as Wallerian degeneration. The myelin sheath surrounding the distal segment becomes disarranged. Debris is removed by macrophages. In the CNS, astrocytes form a glial scar. In the PNS, remaining Schwann cells proliferate to form compact cords. In the CNS and PNS, the proximal severed axons develop sprouts. In the PNS, the delicate axonal sprouts continue to grow until one reaches, and penetrates along, an appropriate Schwann cell cord. The new axon will grow at a rate of 0.5-3.0 microns/day until it reaches and reinnervates its target, i.e., skeletal muscle fiber, sensory receptor in skin, etc. Failure of the sprouting axons to find their Schwann cell cords may result in incomplete regeneration or formation of a painful neuroma. In the CNS, reinnervation is typically very limited due to the sprouting axons being prohibited from reaching their target by the glial scar.

GROSS FEATURES OF THE BRAIN AND BRAINSTEM

I. The adult cerebral cortex, brainstem and cerebellum are divided into the following five regions:

A. Telencephalon - The telencephalon consists of the cerebral cortex, the underlying white matter and a nuclear mass buried deep within each hemisphere (basal ganglia). Convolutions of the human cerebral cortex result in hills and valleys, or more appropriately, gyri and sulci, respectively. The longitudinal cerebral fissure divides the cerebral cortex into right and left hemispheres. Each hemisphere is divided into four lobes by the lateral (Sylvian) fissure, the central sulcus (of Rolando) and the parieto-occipital sulcus. A fifth lobe termed the insula is often overlooked since it can be viewed only if one separates the parietal and temporal opercula adjacent to the lateral fissure.

B. Diencephalon - The cerebral cortex surrounds the lateral and superior aspects of the diencephalon, thus only its inferior aspect can be viewed in an intact specimen. This arrangement enables the diencephalon to be best illustrated in mid-sagittal or coronal sections. Structures of the diencephalon include the epithalamus, thalamus, hypothalamus and subthalamus (see section on DIENCEPHALON for details).

C. Mesencephalon - The mesencephalon, also referred to as the midbrain, contains several clinically relevant nuclei, including the nuclei for cranial nerves III and IV, which exit the ventral and dorsal surfaces of the mesencephalon, respectively. In fact, CN IV is unique in that it is the only cranial nerve to exit the dorsal aspect of the brainstem. Additional important external features of the dorsal aspect of the mesencephalon are the protuberances of the superior and inferior colliculi, which are involved in the visual and auditory pathways, respectively. On the ventral surface, the two large cerebral peduncles are apparent and contain axons travelling from the cortex to inferior regions of the brainstem and spinal cord.

D. Metencephalon - The dorsal aspect of the metencephalon is covered by the cerebellum. Removal of the cerebellum exposes the cerebellar peduncles and the rhomboid fossa, which is actually the fourth ventricle. The pons occupies the rostral half of the rhomboid fossa. On the floor of the rhomboid fossa, the eminence of the facial colliculus is apparent. On the ventral aspect of the pons, numerous neuronal perikarya and axons comprise the prominent bulge termed the basis pontis (basal pons). Cranial nerve nuclei in the metencephalon that project their axons out the ventral aspect of the brainstem include CN V which exits ventro-laterally and cranial nerves VI, VII and VIII which exit in the ponto-medullary groove.

E. Myelencephalon - The myelencephalon consists of the medulla oblongata which begins roughly at the middle of the fourth ventricle dorsally and the ponto-medullary junction ventrally and ends at the level of the first segment of the spinal cord. On the floor of the rhomboid fossa, apparent features are the stria medullaris and the eminences formed by the vagal and hypoglossal nuclei. Caudal to the fourth ventricle, three sulci delineate the cuneate and gracile tuberculi. On the ventral aspect of the medulla, the descending motor fibers form the pyramids and the pyramidal decussation, which are adjacent to, and across, the anterior median fissure, respectively. Immediately lateral to the pyramids are the olives which are eminences formed by the underlying inferior olivary nuclei. Cranial nerve nuclei found within, and project out of, the medulla include CN IX, X and XI which exit the medulla in the postolivary sulcus and CN XII which exits in the sulcus between the pyramids and olivary eminences in the preolivary sulcus.

VASCULATURE AND VENTRICULAR SYSTEMS OF THE CNS

I. Arteries of the CNS - The arterial blood supply of the cerebral cortex, cerebellum and brainstem is derived from branches of the two vertebral and two internal carotid arteries. These two arterial systems provide a collateral circulation by communicating which each other through their branches, thus forming the arterial circle of Willis. Significant branches of these arteries and the regions they supply are illustrated and described below.

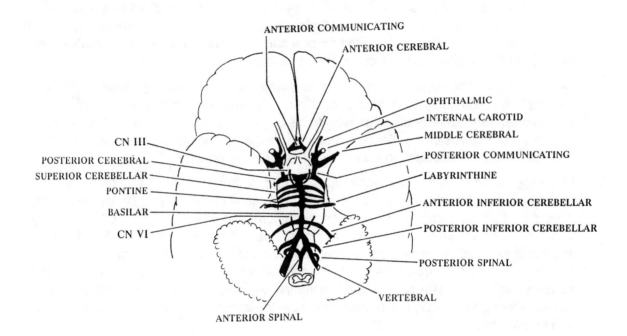

▲ ARTERIAL SUPPLY OF THE CNS

A. Ophthalmic - First branch off of each internal carotid; enters the orbit through the superior orbital fissure; gives rise to the central artery of the retina which pierces the meningeal sheath to travel within the optic nerve and through the optic disc to supply the retina; other branches of the ophthalmic include muscular, lacrimal, ciliary, supratrochlear and supraorbital arteries.

B. Anterior cerebral arteries - Branches of the internal carotid arteries; anterior cerebral arteries course over the corpus callosum in the longitudinal fissure; supply the medial aspects of the cerebral cortices; since motor and sensory function of the lower limbs are represented on the medial aspects of the pre- and postcentral gyri, respectively, occlusion or hemorrhage of the anterior cerebral artery results in deficits in contralateral lower limb function.

C. Anterior communicating artery - Unpaired artery connecting the two anterior cerebral arteries.

D. Middle cerebral arteries - Branches of the internal carotid arteries; the middle cerebral arteries leave the ventral surface of the brainstem through the lateral fissures to supply the

110

dorsolateral surface of the cerebral cortices; realize that occlusion or hemorrhage of the middle cerebral artery may result in deficits in motor and/or sensory function (other than in the lower limbs - see anterior cerebral arteries).

E. Medial and lateral striate arteries - Branches of the middle cerebral arteries that provide the major blood supply to the basal ganglia and internal capsule.

F. Posterior cerebral arteries - Embryologically related to the internal carotid arteries; however, in adults receive their blood from the vertebrobasilar system; supply the occipital, as well as portions of the parietal and temporal lobes of the cerebral cortices; postero-medial and lateral branches of the posterior cerebral arteries supply the thalamus; in terms of the cerebral cortex, occlusion or hemorrhage of the posterior cerebral artery typically results in visual dysfunction.

G. Posterior communicating arteries - Branches of the internal carotids that anastomose with the posterior cerebral arteries.

H. Basilar - Formed by the union of the right and left vertebral arteries; located in a depression on the ventral surface of the pons known as the basilar sulcus.

I. Superior cerebellar arteries - Branches of the basilar artery which course around the brainstem at the pontomesencephalic junction and travel along the superior aspect of the cerebellum; branches from the superior cerebellar arteries supply portions of the pons, lower mesencephalon, superior portions of the cerebellar hemispheres and the cerebellar vermis.

J. Anterior inferior cerebellar arteries (AICA) - The AICA are branches of the basilar artery and course along the pontomedullary junction to supply portions of the pontine tegmentum and inferior cerebellum.

K. Pontine branches - Numerous small branches from the basilar artery that supply portions of the pons and mesencephalon.

L. Labyrinthine arteries - The labyrinthine arteries branch from the basilar artery and course through the internal acoustic meatus to supply CN's VII and VIII; do not contribute to the blood supply of the brainstem.

M. Posterior inferior cerebellar arteries (PICA) - The PICA branch from the vertebral arteries and course around the medulla; important source of blood to the medulla, cerebellar tonsils, inferior cerebellar vermis and inferolateral cerebellar hemispheres; occlusion or hemorrhage of the PICA may result in Wallenberg's syndrome (also called the PICA or lateral medullary syndrome) (see section on MEDULLA).

II. Arterial Blood Supply to the Spinal Cord

A. Posterior spinal arteries - Paired branches from the vertebral arteries that course around, then descend on the posterior aspect of the spinal cord near the dorsal root entry zone; receive contributions from posterior radicular arteries along their descending course; supplies roughly the dorsal one-third of the spinal cord.

B. Anterior spinal artery - Paired branches from the vertebral arteries that fuse into a single anterior spinal artery which descends along the spinal cord in the anterior median fissure; receives contributions from the anterior radicular arteries (which are larger than their posterior counterparts); a significant radicular artery that provides blood to the lumbar enlargement is the great anterior medullary artery (of Adamkiewicz) which is usually present

111

on the left side in the lower thoracic/upper lumbar region; the anterior spinal artery provides blood to the ventral two-thirds of the spinal cord.

III. Ventricular System
The ventricular system is an interconnecting network of cavities within the brain and brainstem. The two lateral ventricles (also referred to as the first and second ventricles, irrespective of side) are present deep to the white matter of the cerebral cortices. Regions of the lateral ventricles, termed horns, correspond with the lobe of the cerebral cortex in which they reside, i.e., frontal, parietal, temporal and occipital. Of particular significance is that the caudate nuclei of the basal ganglia are adjacent to the walls of the lateral ventricles. Specialized tissue known as choroid plexus is present on the floor of the lateral ventricles. The choroid plexus is involved in the production of cerebrospinal fluid (CSF). The CSF produced in the lateral ventricles drains through the foramina of Monro into the unpaired third ventricle. The thalami are adjacent to the lateral walls of the third ventricle. Choroid plexus is also present in the third ventricle. CSF then drains through an opening in the postero-inferior portion of the third ventricle into the cerebral aqueduct. The cerebral aqueduct is small in diameter and descends through the mesencephalon to open into the fourth ventricle. The fourth ventricle is located between the cerebellum and the pons and medulla. The inferior margin of the fourth ventricle is known as the obex. Also, at its inferior margin, the fourth ventricle narrows and continues caudally as the central canal of the spinal cord. Like the other ventricles, choroid plexus is also located in the fourth ventricle. However, the fourth ventricle is unique in that is has two apertures located on each lateral wall (foramina of Luscka) and one located in the midline (foramen of Magendie). The purpose of these foramina is to allow CSF to flow out of the ventricular system into the subarachnoid space. Normally, the choroid plexus produces approximately 600-700 ml of CSF per day. If the flow of CSF through the ventricular system is blocked, the continued production of CSF results in ventricular dilatation (hydrocephalus). Common sites of blockage are the narrow cerebral aqueduct (either by congenital atresia or by a space occupying lesion, e.g., tumor) and the foramina of Monroe.

IV. Meninges

   A. Dura mater - Tough outermost meninx. Dura is divided into a periosteal layer and a meningeal layer. In specific areas, these two layers divide to form the venous sinuses (see below). Also, septa derived from the meningeal layer form the falx cerebri, falx cerebelli, tentorium cerebelli and the diaphragma sella. The dura continues caudally as the spinal dura and is attached inferiorly at the second sacral vertebral level. Stating that something is epidural, i.e., anesthetic block or hematoma, refers to a placement immediately outside of the dura. The subdural region is a potential space.

   B. Arachnoid mater - Non-vascular delicate membrane adjacent, and deep to the dura. In the cranium, the arachnoid does not adhere to the meningeal layer of the dura nor does it extend into the sulci or fissures of the cerebral cortex and cerebellum. In the region of the spinal cord, arachnoid mater does adhere tightly to the dura. Thin finger-like projections, called arachnoid trabeculae extend from the inner surface of the arachnoid to the pia. The arachnoid is suspended off of the pia and neural tissue by the CSF. Regions where the pia and arachnoid are widely separated are known as subarachnoid cisterns. Important cisterns include the:
       1. cisterna magna (cerebellomedullary cistern). CSF drains through the foramina of Luschka and Magendie in the wall of the fourth ventricle into the cisterna magna.

112

Acceptable site for obtaining CSF when not obtainable from the lumbar cistern (see below).

2.   cisterna ambiens (superior cistern). Located around the superior, lateral and posterior margins of the midbrain. Significant because it contains the great vein of Galen (see below) and the posterior cerebral and superior cerebellar arteries.

3.   lumbar cistern. The caudal end of the adult spinal cord with its enveloping pia mater is located at the first and second lumbar vertebral level. The arachnoid, and its overlying dura continue inferiorly to the second sacral level, thus forming the lumbar cistern which contains CSF and the cauda equina. In the adult, the lumbar cistern is the best site for obtaining CSF. However, since the spinal cord extends to lower vertebral levels in the infant, in order to avoid a needle puncture of the spinal cord, an alternate site for obtaining CSF is the cisterna magna.

C.   Pia mater - Vascular membrane that adheres tightly to all components of the CNS and to the dorsal and ventral roots within the vertebral canal. The pia extends into the sulci and fissures of the brain and spinal cord. Denticulate ligaments are specializations of pia found on the lateral aspects of the spinal cord (between the dorsal and ventral roots) along its entire length. These triangular-shaped epipial processes are attached at their base to the pia and at their tip to the arachnoid mater. The function of the denticulate ligaments is to stabilize the spinal cord.

D.   Important features concerning the vasculature and innervation of the cranial meninges - The middle meningeal branch of the maxillary artery provides the major blood supply to the cranial meninges. In addition, the meninges receive anterior meningeal branches from the ophthalmic arteries and posterior meningeal branches from the occipital and vertebral arteries. The larger meningeal arteries are epidural. Thus, an arterial source should be considered when a diagnosis of epidural hematoma is made. Venous blood from the meninges drains into either the diploic veins or the sinuses. The cranial dura is richly innervated by branches from CN V and by branches of the upper two or three cervical spinal nerves.

V. Venous Drainage
In the cranium and orbits, venous blood collects in specialized folds of dura known as sinuses. The sinuses and the direction of blood flow are indicated on the illustration. An important feature of the sinus system is the arachnoid granulations located along the course of the superior sagittal sinus. Arachnoid granulations are protrusions of the arachnoid through the dura into the sinus. They function to facilitate the return of CSF to the venous system.

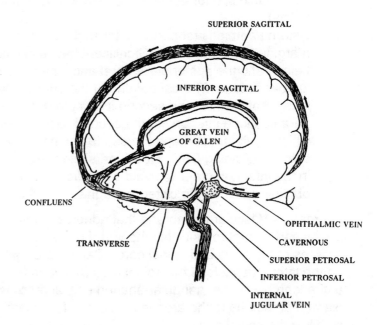

SUPERIOR SAGITTAL
INFERIOR SAGITTAL
GREAT VEIN OF GALEN
CONFLUENS
TRANSVERSE
OPHTHALMIC VEIN
CAVERNOUS
SUPERIOR PETROSAL
INFERIOR PETROSAL
INTERNAL JUGULAR VEIN

▲ VENOUS DRAINAGE FROM THE BRAIN

SPINAL CORD AND PERIPHERAL NERVES

I. The spinal cord

   A. Gross features

      1. Extends from the foramen magnum to the level of the L_2 vertebra in the adult and to approximately the L_4 vertebra in the infant.

      2. Enlargements in the diameter of the cervical and lumbar regions of the spinal cord are due to the large number of motor neurons supplying musculature of the upper and lower limbs, respectively.

      3. The caudal extent of the cord is known as the conus medullaris.

      4. The spinal cord is stabilized laterally by the denticulate ligaments and caudally by the filum terminal.

      5. The ventral surface of the cord is identified by the presence of the ventral median fissure.

      6. Rootlets that join to form the dorsal and ventral roots are attached to the dorsal and ventral surfaces of the cord, respectively.

      7. Lumbar, sacral and coccygeal dorsal and ventral roots contribute to the formation of the cauda equina.

   B. Cross (or transverse) sectional features of the spinal cord

      1. "H"-shaped grey matter surrounded by white matter (the reverse occurs in the cerebral and cerebellar hemispheres).

      2. The upper and lower portions of the "H" are designated dorsal and ventral horns, respectively.

      3. A lateral horn (more commonly referred to as the intermediolateral cell column or IML) is located between T_1 and L_2 spinal cord segments.

a) Realize that because the spinal cord is much shorter than the vertebral column, the location of the IML corresponds to spinal cord segmental levels and not to vertebral levels.

b) ALL sympathetic preganglionic perikarya are located in the IML.

4. A lateral horn is also present between the S_2 to S_4 spinal cord segments.

a) This region, more commonly referred to as the sacral parasympathetic nucleus, contains ALL parasympathetic preganglionic perikarya in the sacral portion of the parasympathetic (cranio-sacral) division of the autonomic nervous system.

b) Realize that the remainder of the parasympathetic preganglionic perikarya are associated with cranial nerves III, VII, IX and X and are located in brainstem nuclei.

c) **Note**: due to the absence of autonomic preganglionic neurons, lateral horns do not exist above T1, between L2 and S2 and below S4.

5. In the 1950's, Rexed divided the grey matter into 10 laminae based on their unique morphological characteristics. Lamina I is located at the tip of the dorsal horn with subsequent laminae named as one moves ventrally through the grey matter (with the exception of lamina X, see below).

a) Lamina I - also termed the marginal nucleus, characterized by the presence of Waldeyer neurons, integration and relay area for terminations of primary afferent axons.

b) Lamina II - also termed the substantia gelatinosa, contains second order neurons and interneurons involved in sensation; important area for relaying sensory information cranially.

c) Laminae III and IV - also termed the laminae propria, receives input terminations from primary afferent axons.

d) Laminae V and VI - located at the base of the dorsal horns, receive and relay sensory information.

e) Lamina VII - also termed the intermediate zone, contains Clark's column which is a relay for the dorsal spinocerebellar tract, contains the IML cell columns in its lateral margin.

f) Lamina VIII - large area located in the ventral horns, neurons in this lamina are interneurons involved in motor pathways.

g) Lamina IX - located in the ventral horn, encircles individual clusters of somatic motor neurons, usually more than one cluster is present.

h) Lamina X - circumscribed area around the central canal, important area for integration of visceral function.

6. White matter surrounding the grey matter is divided into ventral, lateral and dorsal funiculi. Each funiculus contains a large number of myelinated, hence white matter, and unmyelinated axons. Ascending or descending axons that conduct action potentials from similar modalities, i.e., sharp pain, vibration, motor, etc., are grouped into tracts within the funiculi (see below).

II. Tracts

Numerous tracts have been described that link various regions of the spinal cord with higher centers and also transmit different types of information. However, the functions of only 2 sensory and 1 motor tracts are routinely tested in the clinical setting. A concrete understanding of these 3 tracts is paramount to understanding lesions of the spinal cord. The 3 tracts are the spinothalamic tract, the dorsal column/medial lemniscus tract and the corticospinal tract. Each of these is detailed below.

A. Spinothalamic tract
   1. Conducts pain, temperature and crude touch information from body walls and limbs and pain due to distension or ischemia from visceral organs
   2. First order neurons are the pseudounipolar primary afferents with their perikarya in dorsal root ganglia and peripheral processes that project to the distal target.
   3. The central processes enter the dorsal spinal cord and synapse to a large extent on second order neurons in laminae I, II and V of the dorsal horn of the <u>ipsilateral</u> spinal cord. Usually, primary afferent axons have collateral branches that ascend or descend in the tract of Lissauer several segments, thus synapsing on second order neurons 1 to 5 segments from their point of entry into the cord. One reason for this is that this produces an intersegmental dispersion of sensory information.
   4. Primary afferent neurons do not utilize classical neurotransmitters like norepinephrine or acetylcholine. Rather, peptides such as substance P, calcitonin gene-related peptide, vasoactive intestinal peptide, etc., have been implicated in primary afferent neurotransmission.
   5. Axons of the second order neurons cross in the ventral white commissure and ascend in the <u>contralateral</u> lateral spinothalamic (pain and temperature) or ventral spinothalamic (crude and poorly localized touch) tract.
   6. The lateral spinothalamic tract is somatotopically arranged such that inputs from sacral levels are more dorsal in the tract than inputs from cervical levels. The ventral spinothalamic tract is not as somatotopically distinct.
   7. In the lower medulla the lateral and ventral spinothalamic tracts converge to form the spinal lemniscus which ascends through the brainstem to synapse on third order neurons in the ventral posterolateral (VPL) nucleus of the thalamus.
   8. Axons from VPL neurons project as the thalamic radiations through the posterior limb of the internal capsule and terminate somatotopically in the primary sensory cortex (areas 3, 1 and 2). See section on CEREBRUM for details on the cortical representation of sensation.

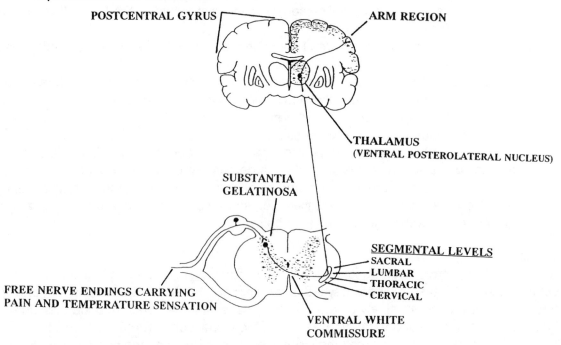

POSTCENTRAL GYRUS

ARM REGION

THALAMUS
(VENTRAL POSTEROLATERAL NUCLEUS)

SUBSTANTIA
GELATINOSA

SEGMENTAL LEVELS
SACRAL
LUMBAR
THORACIC
CERVICAL

FREE NERVE ENDINGS CARRYING
PAIN AND TEMPERATURE SENSATION

VENTRAL WHITE
COMMISSURE

▲ LATERAL SPINOTHALAMIC TRACT

9. During the physical exam, the function of the spinothalamic pathway is assessed by asking the patient to close his/her eyes, then distinguish between a sharp (pin prick) versus dull (small blunt object) placed on numerous dermatomes. In addition, the patient is asked to distinguish between vials filled with hot versus cold water. Crude touch is difficult to assess and its evaluation is not part of the routine physical exam.

10. The most clinically relevant points about the spinothalamic tract to remember are that: 1) it transmits information concerning pain and temperature and 2) it becomes contralateral at, or near, its entry into the spinal cord.

B. Dorsal column/medial lemniscus tract

1. Conducts non-noxious information such as two-point discrimination, vibration, discriminatory touch and proprioception (position sense).

2. First order neurons are the pseudounipolar primary afferents with their perikarya in dorsal root ganglia and peripheral processes that project to the distal target.

3. Central process of the first order neurons enter the medial aspect of the dorsal horn. They neither synapse nor cross where they enter the cord. Rather, the central processes enter the posterior funiculus and ascend to the medulla. Note that these processes are arranged somatotopically in the dorsal column such that fibers from caudal levels are located near the midline in the fasciculus gracilis. A septum in the posterior funiculus appears at about the T6 spinal cord level. Above T6, fibers entering the posterior funiculus arrange lateral to the septum in the fasciculus cuneatus.

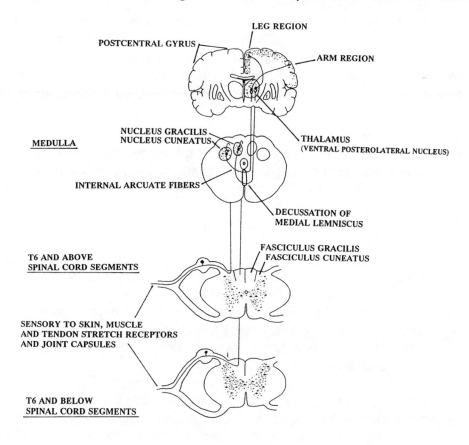

▲ DORSAL COLUMN/MEDIAL LEMNISCUS PATHWAY

117

4. In the lower medulla, first order axons terminate on second order neurons located in the appropriate nucleus gracilis or cuneatus.

5. Axons from the second order neurons project to the contralateral side. This crossing is termed the sensory decussation. The crossed second order axons arrange somatotopically into the medial lemniscus with caudal inputs position more ventrally. In the medulla, the medial lemnisci are positioned in a dorsal/ventral orientation near the midline.

6. In the pons, the medial lemniscus repositions to a more horizontal orientation with the lateral fibers being from the more caudal inputs.

7. Fibers of the medial lemniscus continue to ascend and finally synapse on third order neurons in the ventral posterolateral (VPL) nucleus of the thalamus.

8. Axons from VPL neurons project as the thalamic radiations through the posterior limb of the internal capsule and terminate in the primary sensory cortex (areas 3, 1 and 2). See section on CEREBRUM for details on the cortical representation of sensation.

9. During the physical exam, the function of the dorsal column/medial lemniscus tract is assessed by performing the following tasks over several different dermatomes.
   a) Asking the patient (with eyes closed) to distinguish between a vibrating versus non-vibrating tuning fork.
   b) Asking the patient to describe the position of his/her fingers and toes without looking directly at them.
   c) Brushing the patient lightly with a cotton swab, then asking him/her to state when and where the touch occurred.
   d) Asking the patient to tell you if you touched him/her with one point (using one end of a straightened paper clip) or two (touching with both ends simultaneously).

10. The most clinically relevant points about the dorsal column/medial lemniscus tract to remember are that: 1) it transmits information concerning fine touch, vibration and proprioception and 2) the tract remains ipsilateral throughout the spinal cord; becoming contralateral in the medulla.

C. Corticospinal (pyramidal) tract
   1. Descending primary voluntary motor tract, but not the sole motor tract. Produces finely controlled movements with agility and speed; simple voluntary movements are mediated by other tracts. This tract is facilitory to flexor muscles of the extremities and muscles involved in speech.
   2. The corticospinal tract is composed of axons from pyramidal neurons located in the following areas:
      a) One third from the primary motor cortex (area 4; precentral gyrus). In addition to the pyramidal neurons, Betz cells, known for their unusually large size, also contribute to this portion of the corticospinal tract.
      b) One third from the secondary motor cortex (area 6; precentral gyrus).
      c) One third from the primary sensory cortex (areas 3, 1 and 2; postcentral gyrus). Fibers from the primary sensory cortex are non-motor; rather, they participate in influencing sensory input to the motor system.
   3. Axons from these neurons descend (in order) through the:
      a) corona radiata which is the white matter deep to the cerebral cortex.
      b) genu and posterior portions of the internal capsule.
      c) cerebral peduncles which are located on the ventral aspect of the mesencephalon.
      d) basal pons.

4. In the medulla, approximately 90% of the corticospinal fibers cross as the pryamidal decussation to form the lateral corticospinal tract. The lateral corticospinal tract is somatotopically organized such that axons destined for spinal cord segments innervating the upper limbs are positioned more medially than axons destined for cord segments innervating the lower limbs. The corticospinal tracts are evident on the ventral surface of the medulla as the medullary pyramids.

5. The remaining 10% of corticospinal axons remain uncrossed to form the anterior corticospinal tract. However, fibers in the anterior corticospinal tract usually cross when they reach their destination in the spinal cord.

6. Corticospinal tract axons descend to synapse on interneurons which in turn synapse on alpha motor neurons. A minority of the axons will synapse directly on alpha motor neurons.

7. Approximately 55% of lateral corticospinal tract axons terminate in the cervical region, 20% in the thoracic region and 25% in the lumbosacral regions.

8. Clinically, the patency of this tract is tested by having the patient do a variety of motor tasks that require the use of motor neurons at different levels of the spinal cord.

D. Other significant tracts
1. Corticobulbar tract - Descending projections that supply motor input to various regions of the brainstem, particularly to the motor nuclei of cranial nerves V, VII, IX, XI and XII. Unlike the corticospinal tract that provides motor innervation to contralateral muscles, most bulbar (brainstem) nuclei receive a bilateral innervation from corticobulbar fibers. An important exception to this pattern of innervation is discussed in the clinical correlation section of this chapter.

2. Rubrospinal tract - The rubrospinal tract originates from neurons located in the red nucleus within the mesencephalon. The red nucleus receives inputs primarily from the cerebral cortex and deep cerebellar nuclei. Projections from the red nucleus cross immediately before descending to form the rubrospinal tract which is located in the lateral funiculus. Neurons of the rubrospinal tract form a portion of the extra-pyramidal motor system and function to facilitate contralateral flexor muscles. In addition, they act to inhibit contralateral extensor muscles by hyperpolarizing alpha and gamma motor neurons.

3. Clark's column and the spinocerebellar tracts - This ascending system begins when primary afferent fibers (first order) synapse on a group of spinal cord neurons located in lamina VII, known as Clark's column. Clark's column is present between the C8 and L2 spinal cord segments. The second order neurons in Clark's column ascend in either the dorsal or ventral spinocerebellar tracts. The dorsal spinocerebellar tract ascends uncrossed and enters the cerebellum through the inferior cerebellar peduncle. The ventral spinocerebellar tract, containing the majority of fibers, crosses at its origin in the spinal cord, ascends and then recrosses to the ipsilateral cerebellum through the superior cerebellar peduncle. These tracts function to relay information concerning proprioception, touch, pressure and posture. With this in mind, you should realize that this pathway carries information received via primary afferent processes innervating muscle spindles and Golgi tendon organs.

4. Lateral vestibulospinal tract - Axons of this tract originate from neurons of the lateral vestibular nucleus located in the floor of the fourth ventricle. The vestibular nucleus receives inputs from the vestibular portion of cranial nerve VIII and the cerebellum. The lateral vestibulospinal tract remains uncrossed, descends in the anterior funiculus and functions to maintain balance by facilitating ipsilateral extensor muscles and

inhibiting the activity of ipsilateral flexor muscles. A medial vestibulospinal tract descends primarily to the cervical spinal cord as part of the medial longitudinal fasciculus (in the anterior funiculus) and participates in coordinating head and neck movements relative to eye movements.

5. Tectospinal tract - Axons of the tectospinal tract originate from neurons of the superior colliculus located in the mesencephalon. The superior colliculus receives strong inputs from the visual system, as well as from other types of systems involved in perceiving the environment, e.g., auditory. Axons of the tectospinal tract cross in the mesencephalon and descend in the anterior funiculus as part of the medial longitudinal fasciculus. Tectospinal tract axons terminate on interneurons in upper cervical spinal cord segments and play a role in reflex movements of the head in response to visual and auditory stimuli. Its reciprocal tract, the spinotectal tract, ascends in the lateral funiculus to the superior colliculus, thus completing a reflex "loop".

III. Peripheral nerves

The central nervous system consists of the brain, brainstem, cerebellum and spinal cord. Other components of the nervous system are classified as belonging to the peripheral nervous system (PNS). This section deals with one component of the PNS, namely the spinal nerves. The autonomic nervous system and the cranial nerves complete the PNS and will be dealt with individually in subsequent sections.

A. Formation of spinal nerves

1. Somatic and visceral efferent perikarya have axons that leave the spinal cord via the ventral roots. **Note**: visceral efferent axons, i.e., sympathetic and parasympathetic preganglionic neurons are found only in the T1-L2 and S2-S4 ventral roots, respectively.

2. Somatic and visceral primary afferent perikarya are located in the dorsal root ganglia. **Note**: the only other place where primary afferent perikarya are located is in sensory ganglia associated with the cranial nerves. The central processes of these sensory neurons travel in the dorsal roots and subsequently enter the dorsal aspect of the spinal cord. The peripheral processes of these neurons will travel away from the ganglion toward a somatic or visceral structure.

3. Spinal nerves are formed when the dorsal and ventral roots converge, usually in the intervertebral foramina.

4. There are 31 pairs of spinal nerves: 8 cervical, 12 thoracic, 5 lumbar, 5 sacral and 1 coccygeal.

5. After coursing for just a few centimeters, the spinal nerves diverge and are renamed the dorsal and ventral primary rami.

6. Neurons and individual axons within peripheral nerves are classified according to the type and direction of information they carry. All neurons and axons within peripheral nerves are designated as: General. This designation arises in part from the fact that these axons do not transmit information from a special sense organ (as do axons transmitting visual, olfactory and auditory information). The four classifications or functional components are:

a) General somatic efferent (GSE) neurons - Somatic, in the case of efferent nerves, refers to the innervation of skeletal (striated, voluntary) muscles. The term efferent implies that the action potential is travelling away from the spinal cord, thus these are motor neurons. When one is referring to spinal nerves, the perikarya of GSE neurons are located in the ventral horn of the spinal cord. GSE neurons also supply the striated intrafusal muscle fibers within muscle spindles.

b) General somatic afferent (GSA) neurons - Somatic, in the case of afferent nerves, refers to the innervation of skin, joint and stretch receptors in the muscles and tendons. Afferent implies that these neurons transmit sensory information. For the skin, this information includes pain, temperature, crude touch, vibration, etc. Sensory information from the joints and stretch receptors includes the sense of position (proprioception) and pain (joints only). The sensations of pain, temperature and fine (discriminatory) touch are received primarily by free-nerve endings in the skin and in deeper structures such as tendons and joint capsules. These afferent fibers tend to be small diameter, unmyelinated C-type (slow conducting, burning type pain) fibers or lightly myelinated A delta (fast pain) fibers. These fibers belong to the first order neurons (primary afferents) of the spinothalamic tract pathway. The sensations of being able to discriminate between being touched by one versus two points, vibration, proprioception and light touch are received by encapsulated sensory nerve endings. Examples of encapsulated endings include Pacinian corpuscles, Merkel's corpuscles, Krause end bulbs, Ruffini's corpuscles and Meissner's endings. Nerve fibers associated with these endings typically are of the more heavily myelinated A-alpha and A-beta types.

c) General visceral efferent (GVE) neurons - GVE refers to motor (efferent) neurons of the autonomic nervous system (visceral). These include both the parasympathetic and sympathetic divisions. GVE neurons provide the innervation to smooth muscles and glands and are not under conscious control (e.g., are controlled by structures such as the hypothalamus and brainstem instead of the cerebral cortex). Realize that 2 neurons, a preganglionic and a postganglionic, are required for motor information to travel from the spinal cord to the target organ. Axons from parasympathetic preganglionic neurons, located in the sacral parasympathetic nucleus (S2 to S4 spinal cord segments), travel for a short distance in the S2 to S4 spinal nerves. These nerves are destined for pelvic viscera and the lower colon. Otherwise, parasympathetic axons are not located in any other spinal nerve. However, all spinal nerves contain axons from sympathetic preganglionic neurons. After synapsing in chain ganglia, sympathetic postganglionic axons enter the ventral and dorsal primary rami to innervate blood vessels, sweat glands and erector pili muscles in the skin and blood vessels in deeper structures, i.e., muscles, subcutaneous fat, etc.

d) General visceral afferent (GVA) neurons - GVA neurons carry sensory information from viscera and glands back to areas of the CNS that are involved with regulating involuntary control over body functions. The basic rule is that whatever is innervated by GVE nerve fibers will also be innervated by GVA fibers. Historically, regardless of their intimate relationship, GVA neurons were not included in descriptions of sympathetic and parasympathetic autonomic nerves. Rather, GVA fibers are considered to be "associated" with their respective autonomic counterparts. Indeed, GVA fibers associated with the sympathetic division have their perikarya in the T1 to L2 dorsal root ganglia. Similarly, GVA fibers associated with parasympathetic GVE fibers have their perikarya in the S2 to S4 dorsal root ganglia. One key point to understand is that sympathetic postganglionic GVE axons are found in the ventral and dorsal primary rami, and so to are GVA processes.

IV. Clinical Correlations

A. Upper versus lower motor neuron injury

Manifestations of injury to the motor system, particularly the corticospinal tract, are often described as having upper and/or lower motor neuron characteristics. Unfortunately, students frequently have problems distinguishing which neurons are upper or lower. To simplify this, remember that: 1) lower motor neurons are <u>only</u> the alpha motor neurons and 2) <u>all</u> other neurons in the motor pathway are upper motor neurons. To emphasize this point, alpha motor neurons have their perikarya in the ventral horn of the spinal cord with their axons projecting all the way out to, and synapsing upon, the skeletal muscle fibers. In other words, alpha motor neurons are the final common pathway for all motor innervation. If the alpha motor neuron is damaged, then the muscle fiber will not contract (remember that skeletal muscles must be innervated to contract - this is not the case with cardiac or smooth muscles). On the other hand, if an injury damages interneurons or cortical neurons (upper motor neurons), but leaves the alpha motor neuron intact, muscle contraction is still possible since the muscle is still innervated (recall that the simple reflex arc between primary afferents and alpha motor neurons does not require input from higher neuronal centers to be activated.

1. Deficits resulting from lesions of lower motor neurons:
   a) primary muscle atrophy
   b) hyporeflexia
   c) flaccid paralysis
   d) fasciculations
   Note: a),b) and c) occur due the direct loss of the skeletal muscle innervation; d) is a non-neuronal twitching of muscle fibers.

2. Deficits resulting from lesions of upper motor neurons:
   a) no primary muscle atrophy (secondary muscle atrophy occurs because of disuse)
   b) hyperreflexia
   c) spastic paralysis
   Note: b) occurs because the strength of the reflex arc is no longer dampened (inhibited) by higher neuronal centers.

B. Brown-Sequard syndrome

This syndrome results when the spinal cord is hemisected. For example, a patient received a stab wound in the back that severed the T9 spinal cord segment on the <u>right</u> side. Fortunately, the left side of the spinal cord remained completely intact. The following list details the types of deficits and intact functions you would expect to see.

1. Sensory
   a) Complete anesthesia of the right T9 dermatome. This would occur since the central processes of primary afferent fibers that mediate pain and temperature (spinothalamic tract) and touch, vibration and proprioception (dorsal column/medial lemniscus tract) were destroyed where they entered the cord.
   b) Loss of touch, vibration and proprioception at and below the right T9 dermatome. This would occur since the dorsal column/medial lemniscus tract does not cross until it reaches the medulla. Thus, a right-sided hemisection would transect all ascending dorsal column/medial lemniscus axons on the right side.
   c) Loss of contralateral pain and temperature sense from T9 and below. This would occur since the spinothalamic tract crosses essentially where it enters the cord. Thus, a right-sided hemisection would not only interrupt T9 spinothalamic fibers

that were in the process of crossing, but it would transect all of the second order spinothalamic tract axons that had already crossed (below T9).

    d) Sensations that remain include: 1) pain and temperature below T9 on the right side. This occurs because the second order spinothalamic tract axons below T9 crossed to the non-lesioned side, thus going around the transection and 2) touch, vibration and proprioception on the entire left side. This occurs because all dorsal column/medial lemniscus fibers remained on the left side, never venturing into the area of the lesion.

2. Motor

    a) Lower motor neuron-type lesion of the right T9 dermatome musculature. This occurs because the alpha motor neurons in the right T9 ventral horn were destroyed.

    b) Upper motor neuron-type lesion of muscles on the right side innervated below T9. This occurs because the corticospinal tract crossed in the medulla. Thus, below the T9 level, right sided corticospinal tract axons are destined for motor neurons also on the right side.

    c) Again, since the motor decussation occurs in the medulla, motor function on the left side remains intact.

C. Tabes dorsalis

If left untreated over a period of years, syphilis progresses to a tertiary stage. One sequela of this infection is a degeneration of dorsal column and dorsal root axons, thus interrupting a person's proprioceptive ability. While vibration and touch sense are affected, the deficit in proprioception is the readily apparent as the patient tries to walk. Specifically, the patient has lost his ability to unconsciously know the position of his limbs; therefore in order to walk, he must look at his feet. This behavior, coupled with a deficit in the ability of proprioceptors in his joints to accurately place his feet, results in a clopping, unsteady gait. The dorsal columns are also affected by pernicious anemia with a similar alteration in gait.

D. Amyotrophic lateral sclerosis (ALS; Lou Gehrig's disease)

This is a degenerative disease which results in the loss of alpha motor neurons throughout the spinal cord. Unfortunately, this disease is progressive, without treatment and fatal. Since it affects alpha motor neurons, it is a lower motor neuron disease. Thus, patients with ALS develop progressive signs and symptoms associated with wide spread lower motor neuronal loss, i.e., muscle atrophy, hyporeflexia, fasciculations and flaccid paralysis. Due to the specificity of this disease, there is no impairment in sensory function.

E. Syringomyelia

This disease is caused by a pathological and progressive cavitation of the central canal, primarily in the cervical region of the spinal cord. Progressive widening of the central canal applies pressure on several neural systems which result in a loss of neural function. The first neural system to be affected is the spinothalamic tract. Specifically, the increasing pressure interrupts the axons of second order neurons which cross in the ventral white commissure. This produces a uni- or bilateral loss of pain and temperature. In addition, there is an early and progressive impairment of alpha motor neurons in the ventral horns with a resulting lower motor neuron-type deficit at the affected cord segments. As the cavitation proceeds, axons in the dorsal column may also be affected.

CRANIAL NERVES

In some neuroanatomy courses, a classification scheme is used to describe the various functional components of cranial nerves. Four of these components have been previously discussed, i.e., general somatic afferent (GSA), general somatic efferent (GSE), general visceral afferent (GVA) and general visceral efferent (GVE). In addition to these 4, cranial nerves may possess three others. They are:

> Special visceral afferent (SVA) - sensory neurons related to taste and smell.
> Special visceral efferent (SVE) - motor neurons to skeletal muscles that develop with the branchial arches.
> Special somatic afferent (SSA) - sensory neurons related to vision, hearing and balance.

Realize that a cranial nerve may contain one or more of the seven functional components. However, no cranial nerve contains all seven. The neuroanatomy of the major components of each cranial nerve are described below.

I.   Olfactory nerve - CN I

   A.   The term olfactory nerve is used to identify the unmyelinated axons from bipolar ciliated (hair) cells in the olfactory mucosa. Axons from the olfactory nerves travel through foramina in the cribriform plate, with the majority terminating on mitral neurons within the olfactory bulbs. Cranial nerve I is classified as possessing SVA neurons.

   B.   Axons from mitral neurons project toward the brain as the olfactory tract. Collaterals from these axons synapse in the anterior olfactory nucleus which is a loosely arranged cluster of neurons located within the olfactory tract. Impulses from neurons of the anterior olfactory nucleus cross via the anterior commissure to influence neurons in the contralateral olfactory pathway. At its proximal end, the olfactory tract divides into a large lateral stria, a medial stria and a small intermediate stria.

   C.   Axons in the lateral stria project to areas around the amygdaloid nucleus and prepyriform cortex (region anterior to the amygdaloid nucleus). These regions, located in the temporal lobes, are considered the primary olfactory cortex.

   D.   Axons in the medial stria project to the septal area located on the medial aspect of each frontal lobe.

   E.   Axons of the intermediate stria project to the anterior perforated substance.

   F.   Through multiple pathways, olfactory information is relayed to the hippocampal formation, anterior nucleus of the thalamus, hypothalamus, reticular formation and other structures associated with the limbic system. Through these complex interconnections, odors are intimately associated with regions of the CNS that regulate visceral function, i.e., salivation, nausea, etc.

   G.   Clinical testing of olfaction is accomplished by having the patient smell a common aromatic compound (e.g., coffee) with one nostril closed, then with the contralateral nostril. Types of lesions that affect olfaction include fractures through the cribriform plate and tumors along the route of the olfactory tracts. A complete loss of olfaction is termed agnosia.

   H.   Olfactory "hallucinations", which patients usually characterize as a disagreeable odor, often precede seizure activity involving limbic neurons in the temporal lobe (uncinate fits).

I.   An important point concerning olfaction is that it is the <u>ONLY</u> sensation (somatic, visceral or special) that reaches the cerebral cortex without having been relayed through the thalamus.

II.   Optic nerve - CN II

    A.   Retina

        1.   Cones - Higher threshold of excitability than rods. Responsible for color discrimination and sharp vision. Highest concentration in the fovea centralis region of the retina.

        2.   Rods - Low threshold of excitability. Thus, rods are suited to twilight and night vision.

        3.   Impulses pass from rods and cones through intermediate neurons (bipolar, amacrine, horizontal) that integrate and tune the visual signal. Integrated signals reach ganglion cells whose axons converge at the optic disc to form the optic nerve. Cranial nerve II is classified as possessing SSA neurons.

    B.   Optic nerves converge at the optic chiasm then diverge as the optic tracts. Ganglion cell axons in the temporal retina project in the ipsilateral optic tract. Ganglion cells axons from the nasal retina cross in the chiasm to project in the contralateral optic tract.

    C.   Some ganglion cell axons leave the optic tracts and project to the superior colliculus in the mesencephalon. These axons provide visual inputs to neurons that form the tectospinal tract (see preceding section).

    D.   The majority of axons in the optic tracts terminate in a retinotopic pattern in the lateral geniculate nuclei (LGN) of the thalami. Specifically, each lateral geniculate nucleus is divided into 6 laminae. Uncrossed ganglion cells axons from the ipsilateral eye and crossed axons from the contralateral eye terminate in distinct and non-overlapping lamina within the LGN.

    E.   LGN neurons project posteriorly through the optic radiations to the primary visual cortex (area 17) which is located above and below the calcarine fissure. The primary visual cortex is also retinotopic in that information from lower retinal fields is received in cortex above the calcarine fissure. Conversely, information from upper retinal fields is received in cortex below the calcarine fissure. Information from the primary visual cortex is then relayed to the associational visual cortex (areas 18 and 19).

F. In discussing the visual system, realize that visual fields and retinal fields signify different subjects. For example, light from the left visual fields is projected onto the right retinal fields. Note also that the terms nasal and temporal are used to describe the medial and lateral aspects of both the visual and retinal fields. To emphasize this point, consider that the right nasal retina observes objects in the right temporal visual field. You should understand
this concept for each quadrant of the retina.

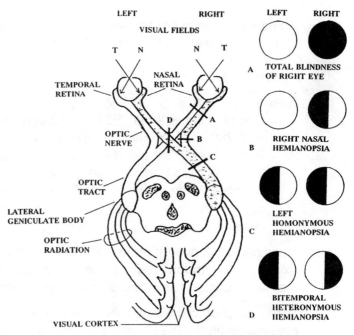

▲ RESULTS OF LESIONS TO THE OPTIC PATHWAY

III. Oculomotor nerve - CN III
CN III emerges from the ventral aspect of the brainstem at the ponto-mesencephalic junction in the interpeduncular fossa. Axons of CN III originate from a compact nuclear complex located near the midline of the mesencephalon and ventral to the periaqueductal grey matter. These nuclei are observed in cross sections of the brainstem obtained at the level of the superior colliculus.

A. GSE neurons are located in the oculomotor nucleus and innervate the following extra-ocular eye muscles (EOM's):
- Superior rectus
- Inferior rectus
- Medial rectus
- Inferior oblique

The levator palpebrae elevates the upper eyelid (opens the eye) and is also innervated by GSE neurons of CN III.

B. Also associated with CN III is the Edinger-Westphal nucleus which gives rise to parasympathetic preganglionic (GVE) axons. These axons terminate on postganglionic neurons located in the ciliary ganglion. Postganglionic axons travel via the short ciliary nerves to the orbit where they innervate the constrictor pupillae muscle. Activation of the Edinger-Westphal nucleus will result in constriction of the pupil.

C. The Edinger-Westphal nuclei receive afferents from the more cranially located pretectal region. Communication between the pretectal region and the Edinger-Westphal nuclei is important in the direct and consensual (indirect) pupillary light reflexes. Specifically, neurons of the pretectal region receive inputs from the visual system. Axons from the pretectum descend to the Edinger-Westphal nuclei as crossed or uncrossed fibers. When light is directed into one eye, the uncrossed fibers activate the ipsilateral Edinger-Westphal nucleus and the ipsilateral pupil will constrict (direct light reflex). Simultaneously, the crossing pretectal fibers will activate the contralateral Edinger-Westphal nucleus resulting in constriction of the contralateral pupil (consensual light reflex).

D. Accommodation-convergence is tested when the patient is asked to visualize the examiners finger when moved from a distant position to a near position. To accomplish this, the patients must:
1. first be able to see the examiners finger (CN II),
2. rotate the eyes medially (medial recti muscles; GSE component of CN III),
3. contract the ciliary muscles which results in a relaxation of the suspensory ligaments, thus causing the lens to become more convex for focusing on near objects (GVE component of CN III) and
4. constrict the pupils (GVE component of CN III).

The pathway for performing accommodation-convergence is believed to proceed from the visual cortex through corticobulbar fibers to the superior colliculus/pretectal region and finally to the oculomotor nuclei. Tertiary syphilis may result in a lesion in this pathway and is characterized by small pupils (miosis) that do not react to light (Argyll-Robertson pupil), but will react to accommodation.

E. Lesions of the oculomotor nuclei or the peripheral portion of CN III result in a lower motor neuron-type lesion characterized by:
1. lateral deviation of the eye (result of the unopposed action of the lateral rectus muscle; CN VI),
2. a fully dilated pupil (mydriasis),
3. ptosis (drooping) of the eyelid (loss of levator palpebrae function) and
4. loss of convergence-accommodation and the pupillary light reflex.

IV. Trochlear nerve - CN IV

A. The trochlear nucleus consists of GSE motor neurons located in sections of the brainstem through the inferior colliculus. Axons leave the nucleus and decussate completely before exiting the dorsal aspect of the brainstem. Recall that all cranial nerves, except CN IV, exit the ventral aspect of the brainstem.

B. The trochlear nuclei innervate the contralateral superior oblique muscles of the eye. Lesions of the nucleus or the nerve may result in vertical diplopia, especially when looking down and to the side contralateral to the lesion (remember, it is a crossed pathway). A dysfunctional superior oblique muscle manifests itself best when patients attempt to walk downstairs.

V. Trigeminal nerve - CN V

A. The trigeminal nerve is the largest cranial nerve. The two types of information transmitted over CN V fibers are:
1. all sensation from the face (GSA) and
2. motor to the muscles of mastication, the tensor tympani, the anterior belly of the digastric muscle and the mylohyoid muscle (first branchial arch derivatives; SVE).

Peripherally, the many branches of CN V organize into ophthalmic, maxillary and mandibular divisions. These 3 divisions converge at the trigeminal (Gasserian, semilunar) ganglion which contains a large number of pseudounipolor GSA neurons. Central processes of these GSA neurons are attached to the ventrolateral pons by a short stump.

B. Sensation from the face can be divided into two types: 1) proprioception and 2) all others, i.e., pain, temperature, touch, vibration, etc.
1. GSA perikarya that mediate proprioception are located, not in the trigeminal ganglion, but in the mesencephalic nucleus of CN V. These GSA neurons are unique in that they

are the only sensory neurons to somatic structures located within the CNS (all others being located in dorsal root or peripheral cranial nerve ganglia). Processes from these neurons travel through the trigeminal ganglion and into the mandibular division to reach muscle spindles and tendon organs located in the muscles of mastication. Collaterals from these axons reach the motor nucleus of CN V to form reflexes associated with the mastication muscles, e.g., jaw jerk reflex.

2. Perikarya associated with all other types of sensation carried by CN V are located in the trigeminal ganglion. Peripheral processes from these neurons project through the three divisions of CN V to terminate as free or encapsulated endings in the skin, oral and nasal surfaces, teeth, cornea and large portions of the cranial dura. The central processes of these neurons enter the pons and descend as the spinal trigeminal tract to synapse on second order neurons in the spinal trigeminal nucleus. Recall that spinal trigeminal tract fibers and second order neurons of the spinal trigeminal nucleus are somatotopically arranged such that the ophthalmic division is located ventrally, the mandibular division is located dorsally and the maxillary division is intermediate in position. Recall that the ophthalmic division of CN V is involved in the blink reflex. Specifically, a cotton swab applied to the cornea activates GSA fibers (ophthalmic division of CN V) which, in turn, activate neurons that innervate the orbicularis oculi muscle (SVE; motor nucleus of CN VII) which closes the eye. Finally, the levator palpebrae muscle (GSE; CN III) is used to reopen the eye. Thus, the blink reflex tests cranial nerves III, V and VII.

3. In addition to projecting to the spinal trigeminal nucleus, collaterals from the central afferent fibers carrying tactile information project to second order neurons located in the principal (main) sensory nucleus which is in the pons, rostral to the spinal trigeminal nucleus. Because of these collateral projections, lesions of the spinal trigeminal tract and nucleus result in deficits in pain and thermal sense, but not tactile sense.

4. Axons from second order neurons of the principal sensory nucleus and the spinal trigeminal nucleus ascend in either the uncrossed dorsal trigeminothalamic or crossed ventral trigeminothalamic tract to synapse on third order neurons located in the ventral posteromedial (VPM) nucleus of the thalamus.

5. Axons from third order neurons in the VPM nucleus project through the posterior limb of the internal capsule and terminate in the primary sensory cortex (postcentral gyrus; areas 3, 1 and 2).

6. An interesting, but devastating disorder of the trigeminal sensory system is trigeminal neuralgia (tic douloureux) which is characterized by excruciating pain, without apparent cause, originating in the dermatome supplied by one or more divisions of CN V.

7. The spinal trigeminal tract and nucleus also serve as the relay for axons from cranial nerves VII, IX and X that innervate small patches of skin around the external ear.

C. The SVE component of CN V originates in the motor nucleus of CN V located in the pons. Axons from this nucleus project through the trigeminal ganglion (without synapsing) into the mandibular division only. Subsequently, these axons terminate on and facilitate the muscles of mastication (i.e., temporalis, masseter and the medial and lateral pterygoid muscles), and the tensor tympani, anterior belly of the digastric and mylohyoid muscles. SVE neurons of the motor nucleus of CN V are controlled by descending corticobulbar projections from the cerebral cortex.

D. Clinical testing of the general function of CN V is achieved by:
1. using a cotton swab to brush the dermatomes of the face supplied by the three divisions of CN V (GSA component of all divisions),

2. corneal blink reflex (GSA component of the ophthalmic division along with cranial nerves III and VII),
3. jaw jerk (GSA and SVE components of the mandibular division) and
4. clinching the teeth and reopening the mouth (SVE component of the mandibular division).

VI. Abducens nerve - CN VI

A. The abducens nucleus contains GSE neurons, is located in the floor of the fourth ventricle, provides the motor innervation to the lateral recti muscles and is encircled on all sides (except ventrally) by the facial nerve. Axons exit the brainstem at the ponto-medullary junction and course forward to the orbit. Two significant points about CN VI are:
1. it has the longest course in the cranium of any cranial nerve and
2. it passes through the cavernous sinus (unlike cranial nerves III, IV and portions of CN V that are embedded in the dural wall of the cavernous sinus). This anatomical relationship with the cavernous sinus makes CN VI vulnerable in the event of thrombi formation or sepsis within the sinus.

B. The abducens nucleus receives crossed and uncrossed inputs from the vestibular nuclei and from the cerebral cortex.

C. A lesion of the abducens <u>nucleus</u> results in a unilateral loss of conjugate lateral gaze. Interestingly, if a lesion occurs to the <u>axons</u> within the brainstem or along their peripheral course, in addition to a loss of lateral gaze, the ipsilateral eye is usually adducted owing to the pull of the intact medial rectus muscle. This is the only case where a lesion of the nucleus versus the nerve produces different symptoms.

VII. Facial nerve - CN VII

A. Cranial nerve VII is composed of two nerves: the facial nerve proper and the nervous intermedius. Although they possess different functional components, the two nerves are usually considered together. In this discussion, we will focus on five of the most important functions of CN VII.

B. SVE - This component arises from the motor nucleus of the facial nerve located in the pontine tegmentum. Interestingly, axons leave the nucleus and course dorsomedially to encircle the abducens nucleus. The fibers then turn ventrally to exit the brainstem in the facial nerve. SVE axons innervate the muscles of facial expression (i.e., orbicularis oculi and oris, zygomaticus major and minor, buccinator, platysma, etc.) and the stapedius muscle which are derived from the second branchial arch.

C. GVE - These visceral efferent neurons are parasympathetic preganglionic in nature. Preganglionic axons arise from neurons located in the superior salivatory nucleus of the pons. Preganglionic axons exit the brainstem in the facial nerve, pass through the sensory geniculate ganglion, then divide into two bundles. One bundle of preganglionic axons travels in the greater petrosal nerve to synapse on postganglionic neurons in the pterygopalatine ganglion. The postganglionic neurons in turn provide the parasympathetic innervation to the lacrimal gland and to glandular tissue located in the nasal cavity and the palate of the oral cavity. The second bundle of fibers descends in the chorda tympani to synapse on postganglionic neurons located in the submandibular ganglion. These postganglionic neurons provide the parasympathetic innervation to the submandibular and sublingual salivary glands.

D. GVA - Visceral afferent fibers originate in glandular tissues supplied by CN VII GVE parasympathetic nerves. This innervation is important for providing sensory information back to autonomic centers in the CNS that regulate glandular function. Afferent processes travel from their targets, through the greater petrosal nerve or chorda tympani to the geniculate ganglion which is located within the petrous portion of the temporal bone. Remember that the geniculate ganglion contains ALL of the sensory pseudounipolar perikarya associated with CN VII (GVA, SVA and GSA). Central processes from these afferent perikarya enter the brainstem and synapse on second order neurons located in the caudal portions of the nucleus tractus solitarius. These second order neurons do not relay visceral afferent information to the thalamus, rather the information is relayed into the medullary reticular formation. The reticular formation is important for regulating autonomic functions.

E. SVA - Taste from the anterior two-thirds of the tongue is mediated via SVA fibers of CN VII. The perikarya of these neurons are located in the geniculate ganglion. Their peripheral processes travel in the chorda tympani which joins the lingual branch of the mandibular division of CN V in order to reach the tongue. Central processes from these neurons enter the brainstem and synapse on second order neurons in the rostral portion of the nucleus tractus solitarius. Unlike the input from GVA fibers, information concerning taste is relayed via second order solitary nucleus neurons to the thalamus. Subsequently, third order thalamic neurons relay taste information to the cortex.

F. GSA - Portions of the external auditory meatus and the skin posterior to the ear are supplied by sensory nerves of CN VII with their perikarya located in the geniculate ganglion. The central processes from these neurons enter the brainstem and synapse upon second order neurons located in the spinal trigeminal nucleus. Sensory information is then relayed rostrally via the trigeminothalamic tract (see CN V).

G. To understand deficits resulting from lesions of the peripheral branches of CN VII, you must know the various functional components each branch contains. For example, a lesion of the right chorda tympani will result in an ipsilateral loss of taste to the anterior two-thirds of the tongue and a decrease in function of the ipsilateral submandibular and sublingual salivary glands.

H. It is important clinically to understand the difference between paralysis of the muscles of facial expression due to Bell's palsy (SVE component of CN VII) versus paralysis due to stroke. Bell's palsy is a condition resulting from a disruption in the function of the motor component of cranial nerve VII. The etiology of this loss of neurotransmission may be due to edema, a virus or unknown, but since it involves the facial motor nucleus in the pons or the facial nerve itself, it is a lower motor neuron lesion. A unilateral Bell's palsy is characterized by paralysis of the ipsilateral muscles of facial expression, both above and below the lateral angle of the eye. A unilateral cerebrovascular accident (CVA; stroke) involving cortical neurons or their axons that innervate the facial motor nucleus is characterized as an upper motor neuron lesion. Interestingly, following an upper motor neuron lesion, one would not expect a paralysis of the facial expression muscles since corticobulbar inputs to brainstem nuclei are bilateral. While this is true for most cranial nerves, only neurons of the facial motor nucleus that supply facial muscles above the lateral angle of the eye receive a bilateral corticobulbar innervation. Thus, following a unilateral CVA, the patient is still able to raise his/her eyebrows and wrinkle his/her forehead, but since facial motor nucleus neurons that innervate muscles below the lateral angle of the eye are innervated only by ipsilateral cortical neurons, the patient experiences paralysis in muscles that control the ipsilateral lower face. In summary, if you detect an ipsilateral paralysis of the muscles of facial expression and the patient <u>cannot</u> raise his/her ipsilateral eyebrows or wrinkle the same side of his/her forehead, think Bell's palsy. If the patient does have function of the muscles above the lateral angle of the eye, think CVA or other lesions that may involve neurons above the facial motor nucleus.

I.   During a routine physical examination, only certain functional components of CN VII are tested. Specifically, taste (SVA) from the anterior two-thirds of the tongue is tested by placing a liquid, such as salty water, on the tip of the tongue and asking the patient to describe the taste. Asking the patient to wrinkle his forehead, smile, etc., tests the SVE component of CN VII. Finally, remember that closure of the eye during the blink reflex is mediated also by the SVE component of CN VII.

VIII. Vestibulocochlear nerve - CN VIII
CN VIII consists of vestibular and cochlear components, both classified as special somatic afferents (SSA). Each will be considered separately.

A.   Vestibular nerve
1.   The vestibular portion of CN VIII is responsible for initiating adjustments of the body relative to gravity, position of the head and to motion. Information from the vestibular nerve is integrated largely in a reflex manner.
2.   True bipolar neurons in the vestibular ganglion, located in the petrous portion of the temporal bone, possess peripheral processes that terminate on hair cells in the crista ampullaris of the semicircular canals and maculae of the utricle and saccule. Changes in angular acceleration and kinetic equilibrium stimulate hair cells in the semicircular canals which in turn initiate an action potential in appropriate fibers of the vestibular nerve. Changes in linear acceleration, gravitational forces on the head and static equilibrium (position of the head in space) stimulate hair cells in the utricle and saccule which in turn initiate an action potential in appropriate vestibular fibers.
3.   Central processes of vestibular ganglion neurons form the vestibular root of CN VIII which is attached to the brainstem at the cerebellopontine angle. Upon entering the brainstem, most primary vestibular fibers terminate on second order neurons in one or more of the four ipsilateral vestibular nuclei. A smaller population of primary vestibular fibers project through the juxtarestiform body (medial portion of the inferior cerebellar peduncle) directly to the ipsilateral cerebellar cortex where they form mossy fiber-type connections (see cerebellum).
4.   Descending projections from the medial and lateral vestibular nuclei are described in the sections on tracts of the spinal cord. Other significant projections from the vestibular nuclei include:
a)   axons that travel via the juxtarestiform body to the cerebellum as mossy fibers,
b)   fibers that ascend in the medial longitudinal fasciculus to synapse on motor nuclei of the extraocular eye muscles (CN III, IV and VI). These fibers are important in most conjugate eye movements and
c)   axons that terminate in the reticular formation. This connection is important in the communication between vestibular input and autonomic function (e.g., sea sickness).
5.   Lesions, irritation or stimulation of the vestibular apparatus may result in vertigo, disturbances in standing or walking, deviations in eye movements, nystagmus and nausea and vomiting.

6. In the unconscious patient, the caloric test takes advantage of the nystagmus reflex. Nystagmus is characterized by a slow movement of the eyes in one direction followed by a rapid, reflex compensation in the opposite direction. The direction of nystagmus is named for the fast component. To perform the caloric test, cold water is placed into either ear, let's say in this case, the right ear. If vestibular function is normal, the eyes will slowly deviate to the right, then rapidly return away from the ear in which the cold water is placed. If warm water is placed in the right ear, the rapid phase of nystagmus will be toward right, or the same ear. Abnormalities in the direction of nystagmus may signify lesions in the central vestibular pathway.

7. An important test in an unconscious patient to assess the extent of damage to the brain and brainstem is to elicit the doll's eye reflex. With the patient lying supine and with the eyes open, the head is rotated left, right, then up and down. If the brainstem neural connections are intact, the eyes in each case will move in the opposite direction in order to keep looking straight. This finding suggest that neural damage may be confined to the cortex. However, if the neural connections between the vestibular nuclei and the nuclei of the extraocular eye muscles are damaged, the eyes will not rotate contra to the head movement. Instead, they will turn with the head or produce disconjugate eye movements. This finding suggests that the damage is more extensive and may involve brainstem nuclei.

B. Cochlear nerve
1. The cochlear portion of CN VIII transmits auditory information. True bipolar neurons located in the spiral ganglion of the inner ear possess a peripheral process that contacts hair cells in the organ of Corti of the cochlea. Central processes of spiral ganglion neurons enter the brainstem, bifurcate and synapse on second order neurons in the ventral and/or dorsal cochlear nuclei located in the pons.

2. Axons from the cochlear nuclei cross in the tegmentum as the dorsal, intermediate and ventral striae. The ventral stria is the largest and is referred to as the trapezoid body.

3. Second order axons may either cross to enter the lateral lemniscus or they may synapse in one of the following ipsi- or contralateral auditory-related nuclei:
    a) nucleus of the trapezoid body,
    b) superior olivary nucleus,
    c) medial accessory nucleus.
    Regardless of where they synapse, projections from these nuclei enter the lateral lemniscus on their respective side.

4. As fibers of the lateral lemniscus ascend, some may synapse in the nucleus of the lateral lemniscus. Neurons from this small nucleus, as well as the remaining lateral lemniscal fibers terminate in the inferior colliculus located in the caudal mesencephalon. It is important to note that information to the nucleus of the lateral lemniscus and the inferior colliculus is relayed to the contralateral side.

5. Neurons of the inferior colliculus project to the medial geniculate body of the thalamus. Subsequently, thalamic neurons project to the primary auditory cortex which is located in the transverse gyrus of Heschl (areas 41 and 42) in the temporal lobe.

6. Note that due to the ipsi- and contralateral projections from the pontine nuclei and relays between the nuclei of the lateral lemniscus and the inferior colliculi, the auditory cortex receives input from both ears. However, this input is predominantly contralateral. Furthermore, a unilateral lesion in the central auditory pathway may result in a bilateral hearing deficit. Not surprisingly, a peripheral lesion involving the cochlear nerve results in a unilateral hearing loss.

7. Clinically, a decrease in auditory acuity must be evaluated to determine if it is due to a conduction or neural deficit (consult an appropriate textbook for an explanation of these tests).

8. Acute loud noises that may be harmful to the ears are dealt with by interesting reflexes. First, the tensor tympani (SVE fibers from CN V) will contract, thus dampening the vibrations of the tympanic membrane. Second, the stapedius muscle (SVE fibers from CN VII) will contract, thus dampening the oscillations of the stapes on the oval window. Finally, motor nerves arising from the superior olivary nucleus project to, and synapse upon, the cochlear hair cells. This pathway, known as Rasmussin's efferent bundle, acts to decrease the stimulatory input from the hair cells to the cochlear nerve.

IX. Glossopharyngeal nerve - CN IX

A. CN IX is a complex nerve that subserves multiple functions in structures of the head and neck. Nuclei associated with CN IX are primarily located in the medulla. Rootlets that converge to form CN IX are located in the post-olivary sulcus of the medulla immediately rostral to the rootlets of CN X. The five most important functions of this nerve are discussed below.

B. SVE - This component originates from a small group of motor neurons in the nucleus ambiguus. Axons project through CN IX to innervate muscles arising from the third branchial arch, i.e., stylopharyngeus and portions of the superior pharyngeal constrictor.

C. GVE - These visceral efferent neurons are parasympathetic preganglionic in nature. Preganglionic axons arise from neurons located in the inferior salivatory nucleus of the medulla. Preganglionic axons travel with CN IX through the jugular foramen, then branch as the lesser petrosal nerve which re-enters the cranial cavity via the tegmen tympani. Preganglionic axons of the lesser petrosal nerve then exit the cranial cavity via the foramen ovale and synapse on postganglionic neurons located in the otic ganglion. Postganglionic axons travel with the auriculotemporal nerve (a branch of the mandibular division of CN V) to innervate the parotid gland.

D. GVA - Visceral afferent perikarya are located in the inferior (petrosal) ganglion which is a noticeable enlargement immediately before the nerve enters the skull. Visceral afferent fibers of CN IX provide unconscious feedback information from the parotid gland. Perhaps more importantly, CN IX provides the sensory innervation to the mucosa of the pharynx, posterior one-third of the tongue, Eustachian tube, tonsils and to baroreceptors located in the carotid sinus. The innervation of the pharyngeal mucosa provides the afferent limb of the "gag" reflex while afferents in the carotid sinus nerve of CN IX relay information concerning arterial pressure. Central process of glossopharyngeal GVA neurons project to the rostral portion of the nucleus tractus solitarius which in turn, relays the information to autonomic-related areas, e.g., medullary reticular formation, to facilitate reflex visceral functions. In addition, collaterals from the sinus nerve project to the dorsal motor nucleus of CN X. An increase in blood pressure stimulates CN IX afferents in the sinus nerve which in turn activates parasympathetic preganglionics of CN X, thus slowing heart rate.

E. SVA - Taste from the posterior one-third of the tongue is mediated via SVA fibers of CN IX. The perikarya of these neurons are located in the inferior ganglion. Central processes from these neurons enter the brainstem and synapse on second order neurons in the rostral portion of the nucleus tractus solitarius. Unlike the input from GVA fibers, information concerning taste is relayed via second order solitary nucleus neurons to the thalamus. Subsequently, third order thalamic neurons relay taste information to the cortex.

F.   GSA - Portions of the external auditory meatus and the skin posterior to the ear are supplied by somatic afferent nerves of CN IX. Unlike other CN IX afferent neurons, GSA perikarya are located in the superior ganglion which is a noticeable swelling along the glossopharyngeal nerve immediately before it emerges from the skull. Central processes from these neurons enter the brainstem and synapse upon second order neurons located in the spinal trigeminal nucleus. Sensory information is then relayed rostrally via the trigeminothalamic tract (see CN V above).

G.   The function of CN IX is tested by touching the back of the pharynx with a tongue depressor. This activates CN IX afferents which in turn stimulate reflex contractions of the pharyngeal constrictors (predominantly innervated by CN X). This disagreeable test is known as the gag reflex. Taste from the posterior one-third of the tongue can also be evaluated (SVA fibers and the nucleus tractus solitarius).

X.   Vagus nerve - CN X

A.   CN X is a complex nerve that subserves multiple functions in the head, neck and viscera within the body cavities. Nuclei associated with CN X are located primarily in the medulla. Rootlets that converge to form CN X are located in the post-olivary sulcus of the medulla immediately caudal to the rootlets of CN IX. The five most important functions of this nerve are discussed below.

B.   SVE - This component originates from motor neurons in the nucleus ambiguus. Axons project through CN X to innervate muscles that develop within the fourth and sixth branchial arches. Specifically, pharyngeal branches (fourth arch) supply the motor innervation to the cranial pharyngeal constrictors and to the levator veli palitini while the superior laryngeal nerve supplies the cricothyroid muscle. The recurrent laryngeal nerve (sixth arch) supplies the motor innervation to the caudal pharyngeal constrictors and the laryngeal muscles.

C.   GVE - These visceral efferent neurons are parasympathetic preganglionic in nature. Preganglionic axons arise from neurons located in the dorsal motor nucleus of the vagus (as well as in the nucleus ambiguus). Preganglionic axons travel with CN X through the jugular foramen, through the neck then synapse on postganglionic perikarya within ganglia located near or within thoracic, abdominal and some pelvic viscera.

D.   GVA - Visceral afferent perikarya are located in the inferior (nodose) ganglion which is a substantial enlargement of the vagus just prior to where it enters the skull. Visceral afferent fibers of CN X provide unconscious feedback information from visceral structures in the body cavities. In addition, CN X provides the sensory innervation to the mucosa of the larynx, root of the tongue and to the carotid body. The innervation of the carotid body is important for relaying information concerning oxygen tension in the blood to the CNS. Central processes of vagal GVA neurons project to the nucleus tractus solitarius. Neurons of the nucleus tractus solitarius relay the information to autonomic-related areas, e.g., medullary reticular formation, to facilitate reflex visceral functions.

E.   SVA - Taste from the epiglottis and root of the tongue is mediated via SVA fibers of CN X. The perikarya of these neurons are located in the inferior ganglion. Central processes from these neurons enter the brainstem and synapse on second order neurons in the nucleus tractus solitarius. Information concerning taste is relayed via second order solitary nucleus neurons to the thalamus. Subsequently, third order thalamic neurons relay taste information to the cortex.

F. GSA - Portions of the tympanic membrane, external auditory meatus and the skin posterior to the ear are supplied by somatic afferent nerves of CN X. Vagal GSA perikarya are located in the small superior ganglion which is located in or beneath the jugular foramen. Central processes from these neurons enter the brainstem and synapse upon second order neurons located in the spinal trigeminal nucleus. Sensory information is then relayed rostrally via the trigeminothalamic tract (see CN V above).

G. The function of CN X is tested by demonstrating that the motor component of the gag reflex is intact (see CN IX, section G) and by having the patient swallow (both tests evaluate the SVE innervation to the pharynx). The SVE component is also tested by observing the uvula while having the patient say "aaahhh". Normally, the uvula should elevate in the midline. If vagal function is impaired unilaterally, the uvula will elevate (and point) away from lesioned side.

H. The vagus provides the parasympathetic innervation to the heart. When the heart rate is abnormally accelerated (as in paroxysmal atrial tachycardia), the vagus can be stimulated to slow the heart rate by rubbing the neck at the carotid bifurcation. This massaging action stimulates the sinus branches innervating the carotid sinus (CN IX) and body (CN X), thus initiating a reflex increase in parasympathetic tone to the heart (reducing heart rate).

XI. Spinal accessory nerve - CN XI

A. CN XI is composed of SVE fibers that originate from cranial and spinal nuclei. The cranial nucleus is composed of neurons in the most caudal segment of the nucleus ambiguus. These neurons give rise to motor axons that leave the brainstem in the post-olivary sulcus. The spinal portion originates from neurons located in the upper cervical spinal cord segments. Axons from these neurons exit the spinal cord as rootlets between the dorsal and ventral roots. The rootlets converge and ascend through the foramen magnum, then leave the cranium through the jugular foramen. Axons from CN XI provide the motor innervation to the ipsilateral sternocleidomastoid and trapezius muscles.

B. A lesion of the central or peripheral components of CN XI results in an ipsilateral lower motor-type deficit. During a physical exam, the spinal portion of CN XI is tested by placing your hand on the patient's mandible, then asking the patient to turn his/her head against resistance. If a lesion is present, weakness will be observed. Realize that weakness observed when turning the head to the right signifies that the left sternocleidomastoid is not functioning properly. The innervation to the trapezius muscles are tested by placing your hands on the patient's shoulders, then asking the patient to shrug his/her shoulders against resistance. In this case, weakness on the right side indicates a decrease in the innervation and thus function of the right trapezius.

XII. Hypoglossal nerve - CN XII

A. The hypoglossal nerve originates from GSE neurons located in a column which is observed in sections through the medulla at the level of the inferior olivary nucleus. Uncrossed axons from this nucleus leave the medulla as rootlets in the pre-olivary sulcus (between the protuberances on the ventral medulla formed by the pyramids and the inferior olivary nucleus). Axons of CN XII innervate all of the intrinsic and most of the extrinsic muscles of the tongue.

B. To test CN XII, the patient is asked to protrude his/her tongue. In the normal individual, the tongue will protrude in the midline. A lesion of the hypoglossal nucleus or nerve results in an ipsilateral lower motor neuron-type lesion. If the lesion occurs on the right side, when the tongue is protruded, it will deviate to the right (toward the lesion) since the right genioglossus muscle is not able to offset the action of the left genioglossus muscle.

XIII. Quick reference to cranial nerve nuclei

A. Nucleus ambiguus - SVE: cranial nerves IX, X and XI

B. Nucleus tractus solitarius - GVA and SVA: cranial nerves VII, IX and X

C. Superior salivatory nucleus - GVE: cranial nerve VII

D. Inferior salivatory nucleus - GVE: cranial nerve IX

E. Spinal trigeminal nucleus - GSA: cranial nerves V, VII, IX and X

MEDULLA OBLONGATA

When asked to identify nuclei within a brainstem slice, it is helpful to be able to recognize the level of the brainstem where the slice was taken. Clues for identifying the medulla, pons and mesencephalon are presented at the beginning of this and the succeeding two sections.

I. How do you know you are in the medulla?

A. Sections through the medulla are almost always taken through either the pyramidal or sensory deccussations or the inferior olivary nucleus. Keys to identifying these slices and important structures within each are listed below.
  1. Pyramidal decussation - Located at the caudal extent of the medulla, thus it appears very much like the spinal cord with its "H"-shaped grey matter and small central canal. However, the ventral median fissure is noticeably absent owing to the crossing corticospinal axons. Also, the inferior extent of the nuclei cuneatus and gracilis are evident in the "posterior funicular" region. **Note**: most tracts occupy the ventral half of the section and that the only cranial nerve nucleus present is the spinal nucleus and tract of CN V.
  2. Sensory decussation - Located <u>rostral</u> to the pyramidal decussation. Important characteristics include the large cuneatus and gracilis nuclei with their crossing internal arcuate fibers and the ventrally located corticospinal axons that form the pyramids. The medullary reticular formation occupies most of the central grey region. Cranial nerve nuclei present in this section include the spinal nucleus and tract of CN V and the hypoglossal nucleus.
  3. Inferior olive - The most striking feature in these sections is the large, convoluted inferior olivary nucleus. Note also that the medial lemniscus occupies a central position. Important cranial nerve nuclei include the tract and nucleus of CN V, the motor nuclei of nerves IX, X and XII and the nucleus tractus solitarius and nucleus ambiguus.

B. Clinical correlation - The most frequent causes of lesions in the medulla are due to demyelinating diseases (e.g., multiple sclerosis), neoplasms and vascular compromise. The demyelinating diseases affect the ascending and descending tracts, thus symptoms are related to the specific tracts involved. Neoplasias can produce a variety of symptoms depending on their location. Similarly, small hemorrhagic or embolic infarctions in medullary vessels can produce site specific deficits. However, the following two syndromes are associated with infarction of larger vessels supplying the medulla.

1. Medial medullary syndrome - This syndrome results following an occlusion or hemorrhage involving the anterior spinal artery or other anterior branches of the vertebral artery. Signs and symptoms of this syndrome include:
   a) contralateral loss of touch, proprioception, etc. (medial lemniscus),
   b) contralateral upper motor neuron deficit (corticospinal fibers in the pyramids) and
   c) ipsilateral lower motor neuron deficit of CN XII (hypoglossal nucleus).

2. Lateral medullary syndrome (PICA syndrome, Wallenberg's syndrome) - This is the most common vascular syndrome involving the medulla and results from an occlusive or hemorrhagic lesion of the posterior inferior cerebellar artery or the vertebral artery. Symptoms of this syndrome include:
   a) ipsilateral loss of pain and temperature to the face (spinal tract and nucleus),
   b) contralateral loss of pain and temperature to the body (spinothalamic tract),
   c) vertigo, nausea, vomiting, nystagmus and deficits in balance (vestibular nuclei),
   d) dysphonia and loss of gag reflex (nucleus ambiguus; SVE of CN IX and X),
   e) ataxia (inferior cerebellar peduncle) and
   f) ipsilateral Horner's syndrome (ipsilateral loss of descending fibers that project to sympathetic preganglionic perikarya).

PONS

I. How do you know you are in the pons?

A. The pons is divided into a ventral half termed the basis pontis and a dorsal half termed the pontine tegmentum.

1. The basis pontis gives the pons its characteristic bulbous appearance ventrally. Structures responsible for this enlargement include the descending corticospinal tracts and the pontine nuclei. The pontine nuclei serve as a massive relay between motor signals from the cerebral cortex to the cerebellum. Specifically, descending corticopontine fibers synapse on neurons in the pontine nuclei. Axons from neurons in the pontine nuclei project transversely to enter the contralateral cerebellum through the massive middle cerebellar peduncles (pontocerebellar fibers).

2. The pontine tegmentum contains the nuclei for CN V though VIII. However, recall that only CN V actually exits the brainstem in the pons; the other cranial nerves exit the brainstem at either the pontomesencephalic or pontomedullary junction. Note that the medial lemniscus moves from its vertically oriented midline position in the medulla to a more horizontal position adjacent to the crossing fibers that form the trapezoid body of CN VIII. The reticular formation in the pontine tegmentum is extensive and is vital for modulating numerous aspects of visceral function (e.g., heart and respiration rates, urinary function, visceral pain, etc.) and levels of consciousness.

3. Located dorsal to the pontine tegmentum are the superior, middle and inferior cerebellar peduncles. These structures transmit the numerous afferent and efferent fibers tracts that allow for communication of the cerebellum with the rest of the CNS (see section on CEREBELLUM).

4. Realize that a large portion of the fourth ventricle is located dorsal to the pons. This is a distinguishing feature between the pons and the small cerebral aqueduct located at the level of the mesencephalon.

B.   As in the medulla, lesions involving the pons are primarily demyelinating, vascular or neoplastic in nature. Lesions of the pontine tegmentum are usually manifested by various deficits in CN V through VIII function (see specific cranial nerves described previously). Lesions involving the basis pontis result in deficits in contralateral motor function. Recall that the corticospinal tract crosses at the pyramidal decussation in the medulla.

MESENCEPHALON

I.   How do you know you are in the mesencephalon?

A.   Slices through the mesencephalon are easily identified by the dorsal protuberances formed by the colliculi and the prominent crus cerebri (cerebral peduncles) located ventrally. Notice also the cerebral aqueduct with its characteristic periaqueductal grey matter. The mesencephalon is usually represented by transverse sections through either the inferior or superior colliculi.

B.   The mesencephalon is divided into three regions.
   1.   The dorsal region is termed the tectum and contains the inferior colliculus, which is a relay in the auditory pathway (see CN VII) and the superior colliculus, which is involved in the visual pathway and is the origin of the tectospinal tract (see CN II).
   2.   The ventral region of the mesencephalon consists of the crus cerebri (basis pedunculi). The crura are large prominences that contain descending corticospinal and corticobulbar axons. Axons within the crura are arranged such that the corticospinal fibers occupy the medial three-fifths. Within this subdivision, axons related to the lower limb innervation are located laterally, axons related to the innervation of the upper limbs are located in an intermediate position and axons related to the innervation of face and neck muscles are located medially. Of the remainder of the crus, the lateral most one-fifth and the medial most one-fifth contain corticopontine axons.
   3.   The intermediate region of the mesencephalon is termed the tegmentum. Nuclei of CN III and IV are located in the mesencephalic tegmentum. Two other prominent nuclei are located in the mesencephalon.
      a)   Red nucleus - This rather large nuclear structure is apparent by its almost circular appearance. Neurons of the red nucleus give rise to the crossed axons of the rubrospinal tract (see sections on SPINAL CORD AND PERIPHERAL NERVES).
      b)   Substantia nigra - This nucleus forms part of the extrapyramidal motor system (i.e., neurons involved in motor function other than those that give rise to the corticospinal tract), is located adjacent to the crus cerebri and is composed of a pars compacta located dorsally and a pars reticularis which is ventrally located. The substantia nigra is so named because of the large number of dopaminergic neurons in the pars compacta that contain the dark pigment melanin. Afferent projections to the substantia nigra originate primarily from the caudate nucleus and putamen of the basal ganglia. These afferents terminate in the pars reticularis which is considered the receptive zone of the nigra. Dendrites from the dopaminergic neurons in the pars compacta (effector zone) arborize in the pars reticularis. A subpopulation of the dopaminergic neurons project to areas of the thalamus. However, perhaps of more clinical interest is the projection of nigral dopaminergic neurons through the internal capsule to the striatum (caudate and putamen nuclei of the basal ganglia; nigrostriatal tract). Degeneration of nigral neurons in this pathway leads to a decrease in dopamine input to the striatum which contributes to symptoms of Parkinson's disease. Although the causative

factor(s) for why nigral neurons degenerate is unclear, some possibilities include genetic, viral and/or environmental toxins. Major symptoms of this progressive motor disorder include:

(1)   tremor (at rest) in hands (described as "pill-rolling" because the tremor looks as if the patient is rolling a pill between his thumb and fingers),

(2)   cogwheel rigidity (jerky instead of smooth muscle movements demonstrable when the arm is contracted against resistance),

(3)   shuffling gait,

(4)   an inability to initiate movements and suddenly stopping while walking, then not being able to resume,

(5)   flattened affect which is manifested by a decrease in facial expressions.

An important point to remember is that in Parkinson's disease, as well as in other extrapyramidal diseases, the tremor is present at rest, but largely disappears if the limb is in motion (the converse is true in diseases involving the cerebellum). Early symptoms of Parkinson's disease can be attenuated by administering L-dopa (a dopamine precursor; available in pill form). However, as the disease progresses, higher doses of L-dopa are necessary with ensuing drug failure being largely due to the administration of toxic doses of L-dopa. Interestingly, addicts who inject heroin contaminated with a chemical whose acronym is MPTP develop a loss of nigral dopaminergic neurons with subsequent Parkinson-like symptoms. While there is no cure for Parkinson's disease, efforts have been made to transplant nigral neurons or cells from the adrenal medulla capable of synthesizing dopamine into the caudate nucleus. These experimental procedures have resulted in minimal to moderate improvements in function in humans.

DIENCEPHALON

I.   General - The diencephalon is the rostral end of the brainstem and is completely surrounded by the cerebral cortex except ventrally. Important boundaries of the diencephalon are that: 1) it is divided into right and left sides by the third ventricle, thus the diencephalon forms the lateral walls of the third ventricle, 2) its anterior limit is the lamina terminalis and foramina of Monro, 3) its posterior limit is the posterior commissure and 4) dorsally it forms the floor of the lateral ventricle. The diencephalon is divided into the following regions.

II.   Epithalamus - The main parts of the epithalamus are the pineal gland, stria medullaris thalami and the habenular nuclei.

   A.   The pineal gland is a single, conical-shaped structure attached to the posterior aspect of the roof of the third ventricle in the midline. In lower vertebrates the pineal is crucial in coordinating reproductive cycles with the length of daylight hours. This coordination is important for reproduction to occur in appropriate times of the year. The function of the pineal in humans is much less clear. However, tumors of the pineal gland in children have resulted in precocious puberty while destruction of the gland has resulted in a delay the onset of puberty. Lesions of the pineal in post-pubertal individuals do not result in apparent deficits. Calcification of the pineal is common in adults and thus makes it visible on X-rays. Lateral displacement of the calcified pineal on X-rays is suggestive of a unilateral space occupying lesion.

B.   The stria medullaris thalami is composed of fibers that originate from the septal nuclei, anterior nucleus of the thalamus and other regions. Fibers of the stria medullaris thalami terminate in the habenular nuclei.

C.   Neurons of the habenular nuclei project to the interpeduncular nucleus within the mesencephalon. The stria medullaris and the habenular nuclei compose a pathway by which visceral and behavioral information (septal and anterior thalamic nuclei) are relayed to midbrain levels.

III.   Thalamus - The right and left thalami are adjacent to the upper one-half of the third ventricle and serve as an important relay for sensory, motor and behavioral functions. Each thalamus contains numerous nuclei whose positions and functions are much too complicated for this review. Therefore, the following discussion is a simplified version of the thalamic organization and is intended only to highlight some important features of this nuclear mass.

A.   Each thalamus is shaped somewhat like an American football with the pointed ends oriented in an anterior/posterior direction. The "Y"-shaped internal medullary lamina, composed of myelinated axons, divides the thalamus into medial and lateral halves with the anterior nucleus (ANT) positioned within the short arms of the "Y". The medial half contains the dorsomedial nucleus (DM). The lateral half of the thalamus is further subdivided into dorsal and ventral tiers. Nuclei in the dorsal tier include the lateral dorsal (LD) and lateral posterior (LP) nuclei and the pulvinar (P). The ventral tier is composed of the ventral anterior (VA), ventral lateral (VL) and ventral posterior nuclei. The ventral posterior nucleus is subdivided still further into the ventral posterolateral (VPL) and ventral posteromedial (VPM) nuclei. Finally, the medial (MGB) and lateral geniculate (LGB) bodies are positioned along the ventro-caudal edge of the thalamus and are referred to as the metathalamus.

B.   If you have the time, a good neuroanatomy book will provide you with many of the connections of the thalamic nuclei. However, since your boards are probably around the corner, I would at least look over the nuclei described below.
   1.   Anterior nucleus:
        a)   receives afferents from the mammillary bodies (mammillothalamic tract) and the fornix,
        b)   projects to the cingulate gyrus,
        c)   component of the limbic system.
   2.   Ventrolateral nucleus:
        a)   receives afferents from the cerebellum, the basal ganglia and the substantia nigra,
        b)   projects to the primary motor cortex (area 4),
        c)   participates in the initiation of movement and control of muscle tone.
   3.   Ventral posteromedial (VPM) nucleus:
        a)   receives afferents from the spinal nucleus of CN V,
        b)   projects to the primary sensory cortex (areas 3, 1 and 2),
        c)   relay for sensation from the face.
   4.   Ventral posterolateral (VPL) nucleus:
        a)   receives all types of sensory information (except from the face) via the dorsal column/medial lemniscus pathway and the spinothalamic tract,
        b)   projects to the primary sensory cortex (areas 3, 1 and 2).
   5.   Medial geniculate body:
        a)   receives afferents from the inferior colliculus and the lateral lemniscus,

b) projects to the primary auditory cortex (areas 41 and 42; transverse gyrus of Heschl),

c) last relay before auditory information reaches the cerebral cortex.

6. Lateral geniculate body:

a) receives afferents from the ipsilateral temporal retina and from the contralateral nasal retina via the optic tract,

b) fibers from the lateral geniculate project as the optic radiations to the primary visual cortex (area 17).

7. Dorsomedial nucleus:

a) interconnections with widespread areas of the cortex,

b) lesions in this nucleus or its connections may alter personality and memory.

IV. Hypothalamus - Nuclei of the hypothalamus are involved in regulating autonomic and endocrine functions. On medial views of the diencephalon, the border between the hypothalamus and the thalamus is demarcated by the hypothalamic sulcus. The right and left hypothalami are composed of several nuclei which are positioned either in the: 1) supraoptic region (above the optic chiasm), 2) tuberal region (above the infundibulum of the pituitary gland) or 3) mammillary region (at and above the mammillary bodies).

A. Supraoptic region - Two important structures in this region are the supraoptic and paraventricular nuclei. The majority of neurons in the supraoptic nucleus synthesize anti-diuretic hormone (ADH; arginine vasopressin) while the majority of neurons in the paraventricular nucleus synthesize oxytocin. Both hormones are transported along axons from these nuclei, through the infundibulum and stored in terminals in the neurohypophysis (posterior pituitary). Increases in blood osmolarity or the presence of gonadal steroids cause the release of ADH or oxytocin, respectively, into the surrounding capillaries. Destruction of ADH producing neurons results in diabetes insipidus which is characterized by polyuria (excessive amounts of urine) and polydipsia (excessive consumption of fluids).

B. Tuberal region - Neurons within tuberal and preoptic regions project axons that terminate in the median eminence. These neurons synthesize a variety of releasing hormones (e.g., adrenocorticotropin releasing hormone, gonadotropin releasing hormone, etc.) which, following an appropriate stimulus, are released into capillaries of the hypophyseal portal system. (Recall that a portal system consists of two capillary beds connected by veins.) The releasing hormones travel through veins to the second capillary bed located in the adenohypophysis (anterior pituitary) where they act on glandular cells to stimulate the release of hormones (e.g., adrenocorticotropin, follicle stimulating hormone, luteinizing hormone, thyroid stimulating hormone, prolactin, melanocyte stimulating hormone and growth hormone) that have a profound affect on numerous visceral functions.

C. Mammillary region - The principal nuclei in this region are the right and left mammillary bodies. These nuclei are observed as protrusions on the ventral surface of the brainstem immediately caudal to the infundibulum. Through their connections, the mammillary bodies are intimately involved in the limbic system. Specifically, there is a reciprocal connection between the mammillary bodies and the tegmental nuclei of the mesencephalic reticular system. The fornix provides a large input to the mammillary bodies from the hippocampus. A major output from the mammillary bodies is to the anterior nucleus of the thalamus, which in turn projects to the cingulate gyrus.

D. Connections of the hypothalamus - The hypothalamus receives its major afferent innervation from areas of the neuraxis involved in visceral functions. For example, the olfactory system, whose stimulation produces numerous visceral responses (e.g., gustatory, nausea, etc.), is linked to the hypothalamus via the medial forebrain bundle which carries fibers from the septal nuclei, the basal olfactory cortex and the periamygdaloid region. The thalamus, frontal cortex, hippocampus and the amygdala (through the stria terminalis) also project to the hypothalamus. Efferent fibers from the hypothalamus project to the thalamus, the reticular formation and to autonomic regions of the spinal cord via the dorsal longitudinal fasciculus.

E. Functional considerations - The hypothalamus, through its influence over the pituitary gland, regulates numerous endocrine functions which control water intake, growth, smooth muscle contraction in the breast and uterus, metabolic rate and fertility. Similarly, visceral functions are regulated by hypothalamic input to autonomic centers. For example, the antero-medial aspect of the hypothalamus influences parasympathetic components while the postero-lateral hypothalamus influences sympathetic tone. Visceral functions regulated by the hypothalamus include heart rate, peripheral vascular tone, blood pressure, gut motility and secretions and urinary bladder control. Other functions, probably mediated by connections of the hypothalamus with cortical and/or limbic structures, include regulation of feeding, satiety, internal temperature, sleep and wake cycles and rage.

V. Subthalamus - The subthalamic nuclei lie ventral to the thalamus and lateral and caudal to the hypothalamus. The major connection of the subthalamic nuclei consists of a reciprocal pathway with the globus pallidus. Lesions of the subthalamus result in ballismus, which is the most violent type of dyskinesia known. A unilateral lesion of the subthalamus results in hemiballism which is characterized by violent, involuntary, flailing motions of the contralateral proximal extremities and limb girdles. Needless to say, since the subthalamus is involved in regulating motor activity, the subthalamus is considered part of the extra-pyramidal motor system.

BASAL GANGLIA

I. General - The basal ganglia consist of a subcortical nuclear mass of telencephalic (not brainstem) origin. The principal parts of the basal ganglia are the caudate, putamen, globus pallidus and claustrum. Some authors include the amygdala as part of the basal ganglia. However, since the amygdala is primarily involved in emotions and behavior, it will be considered with the limbic system.

A. The basal ganglia constitute a major component of the extra-pyramidal motor system. Other components of this system (described earlier) include the red nucleus, substantia nigra, subthalamic nucleus, and the reticular formation. Components of the extra-pyramidal system participate in complex, multi-synaptic pathways that regulate motor function. This concept is best exemplified when one observes the involuntary muscle contractions that follow a lesion to one of the extra-pyramidal nuclei. These involuntary contractions may take the form of tremors or may be much more exaggerated flailing movements. Clinical examples described earlier include Parkinson's disease (substantia nigra and caudate nucleus) and hemiballism (subthalamic nucleus). These examples along with those described below constitute movement disorders termed dyskinesias. Dyskinesias seen with extra-pyramidal lesions occur at rest and interestingly, disappear during sleep.

B. Another type of movement disorder stems from lesions involving the cerebellum. Cerebellar lesions differ from extra-pyramidal lesions in that the involuntary muscle contractions occur, not at rest, but during a purposeful movement, like lifting a coffee cup. Involuntary contractions during a purposeful movement are termed intention tremors.

C. The following terms are used to describe different parts of the basal ganglia:
1. Striatum, Neostriatum = caudate and putamen
2. Corpus striatum = caudate, putamen and globus pallidus
3. Lentiform nucleus = putamen and globus pallidus
4. Paleostriatum = globus pallidus
5. Archistriatum = amygdala

While knowing these terms may seem trivial, striatum is a frequently used term, thus you would probably benefit by knowing exactly what it is. Each of the basal ganglia nuclei are described below.

II. Caudate nucleus

A. This bilateral, elongated nucleus is located adjacent to the floors of the lateral ventricles. The head of the caudate is located rostral to the thalamus while the body and tail arch to assume a dorsolateral position adjacent to the thalamus. The tail ends at the amygdala.

B. Afferent projections to the caudate originate in the cerebral cortex, thalamus and substantia nigra.

C. The principal efferent projections from the caudate are to the globus pallidus and to the pars reticularis of the substantia nigra.

D. The general function of the caudate appears to involve the inhibition of cortically induced motor activity.

III. Putamen

A. The putamen is the largest nucleus of the basal ganglia and is located between the external capsule and the globus pallidus.

B. Afferent input to the putamen originates from the cerebral cortex and the substantia nigra.

C. Efferent projections from the putamen terminate in the globus pallidus and the substantia nigra.

D. The general function of the putamen appears to involve inhibition of cortically induced motor activity.

E. **Special note:** Lesions of the striatum (putamen and caudate) may result in dyskinesias termed athetosis or choreas.
1. Athetosis is characterized by slow, writhing movements particularly of the extremities, but may also involve the face and neck.
2. Choreas are characterized by sudden, involuntary, jerky movements of the distal extremities, face, tongue and deglutitional muscles. The two types of choreas most encountered are:
   a) Sydenham's chorea - This chorea results from rheumatic fever in children. It typically resolves after a short time.
   b) Huntington's chorea - This chorea is an inherited disorder with an onset during midlife. It is characterized by progressive choreiform movements and dementia. Neurochemically, concentrations of the inhibitory neurotransmitter, GABA are reduced in the striatum from patients with Huntington's disease. Conversely,

concentrations of dopamine and tyrosine hydroxylase (the enzyme necessary for catecholamine synthesis) are unaltered. From these and other findings, it appears that the choreiform movements may result from the loss of the inhibitory (GABA) control normally exerted by the striatum.

IV. Globus pallidus

A. The globus pallidus is a bilateral, small nuclear mass that lies medial to the putamen and forms the main efferent fiber system of the basal ganglia. Consider that although the globus pallidus is small, it still contains approximately 650,000 neurons!

B. Afferent projections to the globus pallidus originate primarily from the caudate and putamen (striopallidal fibers) and the subthalamic nucleus.

C. Efferent projections from the globus pallidus are divided into four fiber tracts. The **ansa lenticularis** and the **lenticular fasciculus** utilize different portions of Forel's fields to form the thalamic fasciculus which terminates in several nuclei of the thalamus. The **pallidotegmental fibers** (subthalamic fasciculus) descend to terminate on pontine nuclei. The **pallidosubthalamic fibers** project through the internal capsule to the subthalamus. Likewise, subthalamic fibers project back to the globus pallidus.

D. In experimental animals, bilateral ablation of the globus pallidus results in a noticeable absence of movement giving the appearance of a "sleepy" animal.

V. Claustrum

A. This thin nucleus is positioned between the axons of the external and extreme capsules.

B. The claustrum has reciprocal connections with wide areas of the cerebral cortex, including the visual cortex.

C. The claustrum may participate in convergence of sensory information from various regions of the sensory cortex.

CEREBELLUM

I. The cerebellum is located in the posterior cranial fossa, is covered superiorly by the tentorium cerebelli and is attached to the brainstem by three bilateral peduncles. The cerebellum is divided by the primary and posterolateral fissures into anterior, posterior and flocculonodular lobes. An important feature of the posterior lobe is an area termed the cerebellar tonsils. An increases in intracranial pressure (e.g., hematoma following head trauma) may result in the tonsils being displaced downward (herniated) into the foramen magnum. Herniation of the cerebellar tonsils is a life threatening emergency since blood flow through the vertebral artery system is compromised and damage to the medullary/spinal cord junction is inevitable.

II. The cerebellum is often described along phylogenetic lines.

A. Archicerebellum - This region includes the flocculonodular lobe and is the oldest phylogenetically. The archicerebellum receives numerous inputs from the vestibular system.

B. Paleocerebellum - The anterior lobe and portions of the posterior lobe comprise the paleocerebellum which is more recent phylogenetically than the archicerebellum. These regions receive strong inputs from the spinal cord and help modulate muscle tone, gait and posture.

C.   Neocerebellum - The remainder of the posterior lobe comprises the neocerebellum which is phylogenetically the most recent region. The posterior lobe receives its major input from the pons through the cortico-ponto-cerebellar tract.

III. The outer surface of the cerebellum is covered by a series of elevations termed folia. Cross sections through the cerebellum reveal four pairs of deep nuclei embedded within white matter. The white matter is in turn covered by grey matter. The grey mater is organized into the following layers:

A.   Molecular layer - This outermost layer contains basket and stellate neurons, dendrites from Purkinje neurons and numerous axons.

B.   Purkinje layer - The Purkinje layer is deep, and adjacent to the molecular layer. This layer contains only the perikarya of the Purkinje neurons.

C.   Granule layer - This innermost layer contains a large number of granule and Golgi neurons.

IV. Cytology

A.   Purkinje neurons - Purkinje neurons posses an elaborate dendritic tree that is oriented perpendicular to the long axis of the folia. Each Purkinje neuron has a single axon that projects through the granule layer to synapse on neurons of the deep cerebellar nuclei. However, some Purkinje axons bypass the deep cerebellar nuclei, leave the cerebellum and form direct connections with the lateral vestibular nuclei. Purkinje neurons are inhibitory, utilize GABA as their neurotransmitter and form the main output from the cerebellar cortex.

B.   Basket neurons - Basket neurons receive excitatory input from climbing fibers (see below). Dendrites and axons from basket neurons are confined to the molecular layer. Basket neuron axons are oriented transverse to the long axis of the folia and make synaptic contact with dendrites, perikarya and axons of approximately 10 Purkinje neurons. Basket neurons are inhibitory and utilize GABA as their inhibitory neurotransmitter.

C.   Stellate neurons - These inhibitory neurons are similar to Basket neurons except that their contacts are only to the dendrites of Purkinje neurons. They utilize GABA as their inhibitory neurotransmitter.

D.   Granule neurons - These small neurons are the only excitatory neurons intrinsic to the cerebellum and utilize the excitatory amino acid, glutamate as their neurotransmitter. Their dendrites contribute to the formation of the glomerulus and receive afferent impulses from mossy fibers (see mossy fiber system below). Their axons ascend to the molecular layer where they form a "T" with the lateral axonal branches traveling parallel to the plane of the folia. Granule neuron axons contact Purkinje neurons only once, however a single granule neuron may contact many hundreds of Purkinje neurons along the course of its axonal branches.

E.   Golgi neurons - These large cells are inhibitory and use the inhibitory neurotransmitter GABA. Dendrites from these neurons extend throughout the cerebellar cortex. Axons from these neurons contribute to the formation of the glomeruli in the granule layer.

V.   Afferents to the cerebellum

A.   Many nuclei and regions of the central neuraxis project axons into the cerebellum. Upon entering the cerebellum, these afferent fibers terminate in the cortex as either climbing fibers or mossy fibers.

B. **Climbing fibers** originate primarily from the inferior olivary nuclei in the medulla. These fibers travel through the granule and Purkinje layers to synapse on the dendrites of Purkinje neurons. One climbing fiber will synapse on only one Purkinje neuron, however, it may form from 1 to 2000 synapses upon the Purkinje neuron it contacts. Collateral axonal branches also synapse on the other four neuron types in the cerebellar cortex. The climbing fiber input is excitatory. Since each climbing fiber contacts only one Purkinje neuron, the climbing fiber system is considered to be specific and localized.

C. **Mossy fibers** originate from all of the other afferent sources to the cerebellum. Once in the cerebellum, these fibers bifurcate such that branches may project to different folia. Mossy fibers terminate in the granule layer as glomeruli. A glomerulus consists of a mossy fiber terminal making synaptic contact with Golgi and granule neuron dendrites. In addition, Golgi neuron axons may synapse on the mossy fiber terminal. Mossy fibers provide an excitatory input to the cerebellum and their branching pattern makes them ideal for disseminating diffuse and complex afferent stimuli.

VI. Deep cerebellar nuclei

A. The four pairs of deep cerebellar nuclei are the dentate, emboliform, globose and fastigial.

B. Axons from Purkinje neurons located throughout the cerebellum provide inhibitory input to all of the deep cerebellar nuclei. However, as was mentioned earlier, some Purkinje axons bypass these nuclei to synapse in the lateral vestibular nucleus. The deep cerebellar nuclei also receive direct input from the inferior olivary nuclei, vestibular nuclei and the red nucleus.

C. Axons from the deep cerebellar nuclei project through the inferior cerebellar peduncle to the inferior olivary nuclei, the reticular formation, the red nucleus or the thalamus. Interestingly, while the majority of the intrinsic circuitry of the cerebellum is inhibitory, the output from the deep cerebellar nuclei, hence the cerebellum, is excitatory (see E below).

D. The function of the cerebellum is to coordinate the contractions of skeletal muscles, to maintain proper muscle tone and to participate in maintenance of equilibrium. More specifically, the cerebellum functions to fine tune movement. Output from the deep cerebellar nuclei can be summarized as follows:
   1. projections to the vestibular nuclei influence extensor muscle tone bilaterally,
   2. projections to the red nucleus influence ipsilateral flexor muscle tone,
   3. projections to the reticular formation subserve reticular inhibitory influences and
   4. projections to the thalamus exert their influence indirectly on the contralateral motor cerebral cortex.

E. The neural circuitry within the cerebellum is somewhat paradoxical. For example, when a movement is initiated, the cerebellum acts to coordinate the motion smoothly and precisely. Stated another way, movement increases afferent input to the cerebellum which excites the Purkinje neurons. Purkinje neurons in turn **inhibit** the deep cerebellar nuclei, yet the output from the cerebellum is increased in order to refine the movement. While the reason for this paradox is unclear, other direct inputs to the deep cerebellar nuclei (i.e., from the inferior olivary nucleus) must provide an excitatory source for these nuclei.

VII. All afferent and efferent fibers to the cerebellum must pass through the cerebellar peduncles. The following table lists the peduncle utilized by the major tracts and pathways.

AFFERENTS

Superior cerebellar peduncle
- Ventral spinocerebellar tract
- Tectospinal tract

Middle cerebellar peduncle
- Cortico-ponto-cerebellar fibers (Projections from the cortex to the pontine nuclei are ipsilateral, whereas projections from the pontine nuclei to the cerebellum largely contralateral.)

Inferior cerebellar peduncle
- Olivocerebellar tract
- Dorsal spinocerebellar tract
- Vestibulocerebellar tract
- Trigeminocerebellar tract

EFFERENTS

Superior cerebellar peduncle
- Projections from the cerebellum to the contralateral red nucleus and thalamus (the thalamus, in turn, projects to the motor cortex).

Middle cerebellar peduncle
- Not a significant pathway for cerebellar efferent projections.

Inferior cerebellar peduncle
- Projections from the cerebellum primarily to the ipsilateral vestibular nuclei.

VIII. Clinical correlations

A. Input from stretch receptors (Golgi tendon organs and muscle spindles) is conveyed primarily to the cerebellum. This may appear as a simple responsibility. However, when one considers the simple task of lifting the right foot off of the ground, not only must appropriate flexors in the right leg contract, the extensors in the right leg must be inhibited **and** the contractions of numerous muscles that act on joints in both legs must be coordinated to control balance.

B. The Romberg test is often used to determine proprioception and cerebellar function in the conscious individual. This test is conducted by having the patient stand with his/her feet together and with his/her arms abducted. The patient is then asked to close his/her eyes. Normally, the patient should be able to maintain their position. A positive Romberg test results when balance is not able to be maintained.

C. Cerebellar lesions typically result in ipsilateral defects in movement. To examine cerebellar function during a physical exam, several tests are employed to demonstrate precise motor skills. These test include:
   1. having the patient touch his nose, then move his finger to contact the examiners finger. This motion is repeated several times. If a cerebellar lesion exist, the patient is likely to miscalculate the distance between his nose and your finger (dysmetria; past-pointing).
   2. having the patient rapidly pronate and supinate his hands. An inability to perform this task is known as dysdiadochokinesia.

D.   Other clinical characteristics of cerebellar lesions include intention tremors, disorders of phonation due to deficits in laryngeal muscle function, gait disturbances and nystagmus.

CEREBRUM

I.   The cerebrum consists of an outer cortex that is composed of grey matter, an expansive white matter that underlies the cortical grey matter and several nuclear masses (e.g., basal ganglia) that occupy the deep portions of the cerebrum. The cerebrum is divided into right and left hemispheres by the longitudinal cerebral fissure. Each hemisphere is further divided into five lobes.

A.   Frontal - The frontal lobe extends from the anterior tip of the cortex to the central sulcus (of Rolando). In general, the anterior half (prefrontal area) is concerned with "higher" functions such as personality, interpersonal skills and concepts of person, place and time. The posterior half of the frontal cortex subserves various motor functions.

B.   Parietal - The parietal lobe extends from the central sulcus posteriorly to the parieto-occipital sulcus. It is primarily concerned with receiving, integrating and interpreting sensations.

C.   Occipital - The occipital lobe is bounded anteriorly by the parieto-occipital sulcus and is concerned with receiving and interpreting visual stimuli.

D.   Temporal - The temporal lobe comprises the infero-lateral portion of the cerebrum and is separated from the frontal and parietal lobes by the lateral (Sylvian) fissure. The temporal lobe is intimately linked with, and even contains, limbic structures. Thus, aspects of behavior, memory and "visceral" feelings are synthesized in the temporal lobe. In addition, the temporal lobe contains cortical regions subserving hearing and taste.

E.   Insula - The insula can be observed by retracting the cortex on either side of the lateral fissure. Stimulation of the insular cortex results in visceral sensations and responses.

II.   White matter within the cerebrum

A.   There are an estimated 14 billion neurons in the cerebral cortex whose axons form tracts that project either within or out of the cerebrum. Axons within these white matter tracts can be characterized as follows.
   1.   Associational - These axons project from one area of the cerebral cortex to another area within the same hemisphere. Short associational fibers project from one gyrus to the adjacent gyrus. Long associational fibers form connections between lobes.
   2.   Commissural - These axons project from regions in one cerebral hemisphere to the contralateral hemisphere. Examples include fibers of the anterior and posterior commissures and the massive commissural bundle termed the corpus callosum.
   3.   Projection - These axons leave the cerebrum and project to numerous brainstem nuclei and the spinal cord. Examples include corticobulbar, corticospinal and corticopontine axons.

III.   Histology of the cerebral grey matter

A.   Cross sections through the cortex reveal that it is a laminated structure composed of alternating bands of neurons and fibers. However, the cortex is not uniform in terms of its laminated appearance. For example, some regions have six laminae while others have only three. The following terms are applied to cortical regions displaying certain laminar arrangements.

1. Neocortex - These cortical areas contain six laminae and comprise approximately 90% of the cortex. Neocortex is phylogenetically the youngest and is found in primary sensory and motor areas.
2. Allocortex - This cortical arrangement is composed of three layers. Allocortex is phylogenetically the oldest and is found in areas subserving limbic-type functions, i.e., hippocampus and olfactory cortex.
3. Mesocortex - The laminar arrangement of mesocortex falls in between the neo- and allocortex. An example of mesocortex is found in the cingulate gyrus.

B. Cytology
1. The predominant neuron in the cerebral cortex is the pyramidal neuron. These neurons have several dendrites that course laterally or toward the cortical surface and an axon that travels in a direction opposite the cortical surface. Pyramidal neurons are located in most laminae and form the major projections from the cortex. The giant cells of Betz, which are located in lamina V of the primary motor cortex, are atypical pyramidal neurons because of their relatively large size.
2. Stellate neurons are found in most laminae of the cerebral cortex. These neurons possess dendrites that course in all directions and an axon that may terminate close to the perikaryon or travel to adjacent cortical regions. Stellate neurons appear to act as interneurons in the cerebral circuitry.
3. Fusiform neurons are located in deeper areas of the cortex and are oriented perpendicular to the cortical surface.
4. The horizontal cells of Cajal are located superficially in the cortex and are oriented parallel to the cortical surface.
5. The cells of Martinotti are located in all cortical laminae and send their axons toward the surface of the cortex.

IV. Functional considerations

A. The primary inputs to cortical neurons are from thalamic nuclei and from associational and commissural fibers. Many thousands of synaptic contacts may be present on a single cortical neuron.

B. In Nissl stained preparations, the cerebral cortex has a well delineated laminated appearance that is horizontal to the cortical surface. However, experimental studies have demonstrated that the functional unit of the cortex is not an horizontally arranged group of neurons. Rather, functional units are arranged in vertical columns of neurons extending across all laminae. For example, neurons in a vertical column are all activated by the same type of stimulus applied to the same receptive field. Neurons outside of the column do not respond to the stimulus.

V. Cortical mapping

A. Although the functional unit of the cortex is the cortical column, it is clear that columns that subserve similar functions (e.g., vision, somatic sensation) are located in relatively specific regions on the cerebral cortex. In 1909 Brodmann divided the cortex into 47 areas based on distinctions in their cytoarchitecture. Interestingly, recent physiological studies have determined that Brodmann's areas also correlate quite well with respect to specific functions.

B. Several regions of the cortex that are function specific can be subdivided into a **primary** area and secondary or **associational** areas. These two areas subserve different functions.

Generally, primary areas concerned with sensation (somatic, visual, auditory, etc.) are considered to be the first cortical site where the sensation is received. **Primary** areas are concerned with integration and discrimination of sensory information. A lesion in a primary cortical area usually results in a complete deficit of the particular sense, i.e., lesioning the primary visual cortex results in blindness. A lesion in the primary motor cortex results in paralysis. **Associational** areas receive input from the primary areas, as well as from other areas of the cortex. **Associational** areas act to further integrate and to refine the sensory stimulus. Lesioning associational areas usually does not result in a complete deficit. Rather associational lesions result in an impairment of function, i.e., lesions of the associational visual cortex may result in visual hallucinations and/or visual agnosia.

C. The following list details some of the major primary and associational areas of the cerebral cortex, their Brodmann's number, location and function.

1. Primary sensory (somesthetic) cortex is located in the postcentral gyrus (areas 3, 1 and 2) of the parietal lobe. It receives afferents from the VPL and VPM nuclei of the thalamus in a somatotopic manner. This somatotopy is depicted by the **sensory homunculus**. Lesions result in a site specific loss of sensation. For example, occlusion of the anterior cerebral artery which supplies midline structures results in a sensory loss in the lower limb.

2. Primary visual cortex (area 17) includes the cuneus and lingual gyri located above and below the calcarine fissure in the occipital lobe. It receives afferents from the ipsilateral geniculate body of the thalamus. Lesions result in blindness.

3. Associational visual cortex (areas 18 and 19) is located in the occipital lobe adjacent to the primary visual cortex. Lesions result in visual hallucinations and agnosia.

4. Primary auditory cortex (areas 41 and 42) is located in the temporal lobe within the lateral fissure (transverse gyri of Heschl). It receives afferents from the medial geniculate body of the thalamus. Each primary auditory cortex receives information from both ears. Thus, a unilateral lesion results in a decrease (mostly contralateral) in auditory acuity, but not deafness.

5. Associational auditory cortex (areas 22 and 23) borders the primary cortex. Lesioning this area in the dominant hemisphere results in Wernicke's aphasia. This aphasia is characterized by unimpaired hearing, yet the patient is unable to comprehend sounds, especially speech.

6. Primary gustatory cortex (area 43) is located in the parietal operculum. It receives afferents from the VPM nucleus of the thalamus. Lesioning results in a decrease in discriminative taste. However, crude taste may still be mediated at the thalamic level.

7. Primary olfactory cortex is located near the tip of the temporal lobe in the periamygdaloid nucleus and prepyriform cortex (region anterior to the amygdaloid nucleus). These regions are considered the primary olfactory cortex.

8. Primary motor cortex (area 4) is located in the precentral gyrus of the frontal lobe and contributes approximately one-third of the axons to the corticospinal tract. Neurons in this region are involved in highly skilled volitional movements. The somatotopic pattern of innervation by the primary motor cortex is illustrated in **the motor homunculus**. Seizure activity that originates in one area of the motor cortex (thus producing convulsions of a specific muscle group) and spreads to involve other areas is termed a Jacksonian march seizure. Lesions of the primary motor cortex in humans generally produce a site-specific contralateral paralysis, i.e., fingers, arm, leg, face, etc. However, since the premotor cortex constitutes only a portion of the corticospinal tract, some restitution of function, particularly in the proximal limbs, may occur over time.

9. Premotor cortex (area 6) is located in front of the primary motor cortex. Approximately one-third of corticospinal tract axons arise from this region and electrical stimulation will produce contralateral muscle contractions. Lesions of area 6 alone do not produce the severe muscle paralysis observed following lesions to the primary motor cortex.

10. Frontal eye field (area 8) is located anterior to the premotor cortex. Stimulation of this area produces strong conjugate eye movements to the contralateral side. This area is believed to function in volitional eye movements.

11. Broca's area (areas 44 and 45) is located in the opercular area of the frontal lobe. This area corresponds to the motor speech area in the dominant hemisphere. Lesions of Broca's area on the dominant side result in expressive aphasia. Lesions in the non-dominant hemisphere do not interfere with speech. **Note:** Cerebral dominance is described clinically according to which hemisphere contains the motor speech center. In right-handed individuals, the speech center is almost always located in the left hemisphere. Interestingly, the motor speech area is still in the left hemisphere in approximately **80%** of left-handed individuals, with the remainder having their speech center in the right hemisphere.

12. Prefrontal cortex (areas 9-11) is located in the anterior aspect of the frontal lobe. This area is phylogenetically quite recent and participates in affective judgement and behavior. Beginning in the 1940's a psychosurgical procedure, known as a prefrontal lobotomy (or leucotomy), was utilized to modify the behavior of severely psychotic patients. This procedure involved the bilateral transection of fibers to and from the prefrontal cortex. Such surgeries have been largely replaced by the advent of more effective pharmacologic therapies. Patients with tumors arising in the anterior cranial fossa that impinge on the prefrontal cortex may present with changes in personality (noticed by family members, friends, etc.).

13. Major associational cortex (areas 39 and 40) is located in the supramarginal and angular gyri at the posterior margin of the lateral fissure. This area is connected to all cortical sensory areas and functions in higher order, multisensory perceptions. Lesions of this area may result in:
    a) receptive and/or expressive aphasia (see above),
    b) astereognosis (inability to recognize objects by feel alone),
    c) acalculia (an inability to perform mathematic computations),
    d) constructional apraxia (difficulty in drawing circles or squares) and
    e) disturbances in the recognition of self (shaving the left side of the face only because of an unawareness of the contralateral side).

LIMBIC SYSTEM

I.   To many students the limbic system represents all that is confusing about neuroanatomy. One reason for this confusion is that the limbic system participates in intangible concepts instead of in something more concrete, like perceiving a pin prick or wiggling your toe. For now, let's just say that your limbic system is simply a relay between life as your neocortex experiences it and the physiological manner in which your hypothalamus deals with it by doling out the necessary hormones and autonomic reactions. These hormones and autonomic reactions are what determines emotions, behavior and the integration of responses such as fight or flight, obtaining food, sexuality and motivation.

II.   Limbic lobe - The limbic lobe and the limbic system describe two different things. The limbic lobe consists of the grey matter that is adjacent to, and surrounds the diencephalon. The four parts of the limbic lobe are the:

   A.   cingulate gyrus, which is located immediately above the corpus callosum;

   B.   subcallosal gyrus, which is the anterior and inferior continuation of the cingulate gyrus;

   C.   parahippocampal gyrus, which occupies the medial gyri of the temporal lobe and the

   D.   hippocampus and dentate gyri, which underlie the parahippocampal gyrus.

III. Limbic system - The limbic system is a global term which includes the components of the limbic lobe described above as well as the following associated areas:

   A.   septal nuclei,

   B.   amygdala,

   C.   mammillary bodies of the hypothalamus,

   D.   anterior nucleus of the thalamus,

   E.   epithalamus and the

   F.   olfactory system.

IV.   Three areas of the limbic system not covered in previous sections include the hippocampal formation, amygdala and septal area.

   A.   Hippocampal formation - The hippocampal formation is an infolding of the parahippocampal gyrus in the temporal lobe. The two main components of the formation are the hippocampus and the dentate gyrus (the subiculum is also included in the formation). Neurons in the prepyriform cortex project to pyramidal neurons in the entorhinal cortex. Axons from entorhinal neurons project to pyramidal neurons of the hippocampus via a medial alvear pathway and a lateral perforant pathway. Granule neurons in the dentate gyrus also project to pyramidal neurons of the hippocampus. Axons from the hippocampus project through the fimbria into the fornix where they project to their primary termination in the mammillary bodies (some axons also project to the septal nuclei). From the mammillary bodies, information is relayed through the anterior nucleus of the thalamus to the cingulate gyrus. While a unilateral ablation of the hippocampus does not result in deficits, a bilateral ablation results in a loss of short term memory and an inability to store new information.

B. Amygdala - The bilateral amygdaloid nuclei are located in the temporal lobe immediately anterior to the hippocampus. Each amygdala receives strong afferent input from the olfactory stria, hypothalamus and has reciprocal connections with the frontal lobe. Other efferents from the amygdala project to the thalamus. Under normal conditions the amygdala participates in autonomic and stress responses such as modulation of blood pressure, heart rate, respiration rate and gastric motility. Bilateral lesions of the temporal lobes and amygdala may result in the Kluver-Bucy syndrome. This syndrome is characterized by patients who develop an aberrant tendency to examine almost anything orally, visually and tactually. In addition, patients (males in particular) develop hypersexual tendencies toward any gender. While this syndrome is commonly described for lower species, it has been reported in humans.

C. Septal nuclei - The septal nuclei are located in the infero-medial portion of the frontal lobes. They receive afferents from the hypothalamus and midbrain through the medial forebrain bundle, from the amygdala via the diagonal band of Broca and from the hippocampus via the fornix. Efferents from the septal nuclei project to the hippocampus, hypothalamus, midbrain and habenular nuclei. The function of the septal nuclei is to provide a site of interaction between limbic and diencephalic structures. Physiological studies implicate the septal nuclei in modulating arousal, learning, emotions and sexual behavior.

V. Additional clinical considerations

A. Temporal lobe epilepsy - Limbic structures, particularly the amygdala and hippocampus, have an exceptionally low threshold for seizure activity and tolerance to ischemic insults. Because of these characteristics, seizure activity originating in the temporal lobe is common. Temporal lobe epilepsy, also referred to as psychomotor seizures or uncinate fits, is characterized by olfactory, visual and/or auditory hallucinations, rhythmic facial movements resembling feeding, amnesia of hours to days and performing violent, criminal acts during the seizure without being aware of, or remembering, the act itself.

B. Korsakoff's syndrome - This syndrome is a later development of Wernicke's encephalopathy. The etiology of this syndrome is believed to be a thiamine deficiency secondary to chronic alcoholism. The syndrome is characterized pathologically by a destruction of the mammillary bodies with possible involvement of the thalamus. Korsakoff's syndrome is characterized by a profound deficit in memory, confusion and confabulation.

AUTONOMIC NERVOUS SYSTEM

I. The nervous system is divided into two distinct, but interconnected systems. The first is the central nervous system (CNS) which consists of the brain and spinal cord. The other is the peripheral nervous system (PNS) which can be subdivided into 3 parts. The first part of the PNS encompasses the cranial nerves (CN) which are numbered I through XII. The second part of the PNS includes the 31 pairs of spinal nerves. Spinal nerves branch to form the dorsal primary rami that innervate the skin and muscles of the back and ventral primary rami that innervate similar structures on the lateral and anterior body walls and in the limbs. The last part of the PNS consists of the autonomic nervous system (ANS). The ANS is further subdivided into a sympathetic division and a parasympathetic division.

II. GVE and GVA fibers are associated with the ANS. GVE fibers provide motor function to the smooth muscles of hollow organs (stomach, intestines, blood vessels, etc.), sweat glands and erector pili muscles (the little muscles that cause goose bumps). Since you have sympathetic and parasympathetic divisions of the ANS, you will have both parasympathetic and sympathetic GVE's. Unfortunately, in the early days, GVE's were considered as the only component of the ANS. Sensory nerves traveling with the motor fibers were not included in the original classification of the ANS. In reality, where there is a GVE, there will also be a GVA fiber. In other words, sympathetic GVE nerves also have GVA fibers carrying unconscious sensation from viscera back to the CNS (the same goes for the parasympathetic division). Thus, the afferent limb of the autonomic innervation will be referred to as either: GVA's associated with the sympathetics or GVA's associated with the parasympathetics. One final important point to remember is that only the sympathetic division of the ANS is present in the limbs and body walls. The parasympathetic portion of the ANS is confined to viscera in the head, thorax, abdomen and pelvis. Parasympathetic nerves do not, and will not ever innervate visceral structures (like blood vessels and sweat glands) in the body walls or limbs.

III. In the autonomic nerves to be discussed, remember that we are referring only to visceral efferent (GVE) and visceral afferent (GVA) fibers. Also remember that GVA neurons project all the way from their target to the CNS. In other words, they only synapse at their axon terminals, no where else. A GVE neuron on the other hand, must synapse on a second neuron before reaching its target. In peripheral nerves, <u>THE GVE SYSTEM IS A TWO NEURON SYSTEM AND IT IS THE ONLY TWO NEURON SYSTEM!!!!</u> The first neuron, which originates within the brainstem or spinal cord is termed the preganglionic neuron. The second neuron, which is located in a peripheral ganglion (a collection of perikarya outside the CNS) is termed the postganglionic neuron.

IV. Organization of sympathetic GVE neurons

   A. The sympathetic division is also described as the thoracolumbar division. This term is derived from the location of the sympathetic preganglionic GVE cell bodies. Sympathetic preganglionic cell bodies are located between T1 and L2 in the IML of the spinal cord. Remember that ALL preganglionic sympathetic GVE cell bodies are located in the IML between T1 and L2!! The axons from these neurons project through the ventral roots and through the white ramus communicans into the sympathetic chain ganglia. Once in the ganglion, the preganglionic axons can do one of four things:
   1. Sweat glands, blood vessels and erector pili muscles in the body wall and limbs are innervated by sympathetic GVE's and their associated GVA's. Since there are no ganglia in these places, GVE neurons must synapse before entering the ventral or dorsal primary rami. Therefore, sympathetic preganglionic GVE fibers entering the sympathetic chain ganglia will synapse on a postganglionic neuron, which will then project its axon through the grey ramus communicans into the ventral or dorsal primary ramus. Examples of this type of pathway include the pathway by which sympathetic postganglionic fibers get into intercostal nerves, the ilioinguinal nerve, the ulnar nerve, etc.
   2. Another route for sympathetic GVE's is that the preganglionic fiber entering the chain ganglion may travel to another segmental level. In this case, the preganglionic axon will enter the sympathetic chain and ascend or descend to re-enter, and synapse upon, a sympathetic postganglionic GVE neuron within a chain ganglion at another level. The postganglionic axon will then exit through the grey ramus to enter the spinal nerve. This is rather straight forward if changing levels occurs between T_1 and L_2. However, realize that since the contribution of preganglionic fibers into the chain ganglia is confined to the T_1 to L_2 spinal cord segmental levels, white rami communicantes will

only be found between T_1 and L_2. However, since the sympathetic chain and ganglia extend from the upper cervical to coccygeal <u>vertebral</u> levels and since postganglionic sympathetic GVE fibers must be able to get into all of the spinal nerves, grey rami must be present to connect the chain ganglia with the spinal nerves at all levels. Examples of this pathway include how sympathetic postganglionic GVE fibers enter the cervical plexus which innervates the neck and the brachial and lumbosacral plexi which innervate the arms and legs, respectively.

3. The third route for sympathetic GVE's is that the preganglionic axons enter the chain ganglia and synapse on postganglionic neurons. The postganglionic axon will then leave the ganglion (not through the chain or ventral or dorsal primary rami) and head for an organ. An example of this pathway occurs in the thorax where the cardiac nerves (postganglionic fibers) leave the chain ganglia, travel through the cardiac ganglionated plexus (without synapsing, since they are already postganglionic) to innervate the heart and lungs.

4. The fourth possible route for sympathetic GVE's is that the preganglionic fiber enters, then leaves the chain ganglion without synapsing. Examples of this pathway include the thoracic, lumbar and sacral splanchnic nerves. Remember that even though the preganglionic fibers have exited the sympathetic chain ganglion, they still must find a postganglionic neuron prior to entering an organ. Postganglionic neurons are of course, located in preaortic (prevertebral) ganglia. Once the preganglionic fiber has synapsed on the postganglionic neuron, the postganglionic axon will "catch a ride" on a blood vessel headed for an abdominal or pelvic organ. An example of this would be the innervation of the duodenum. The duodenal innervation starts with preganglionic neurons located between T_1 and L_2 (probably around T_5). The preganglionic axons pass through the chain ganglion (without synapsing), then enter the greater thoracic splanchnic nerve. The greater splanchnic nerve then descends through the diaphragm whereby its axons synapse on postganglionic neurons located in the celiac ganglion. The postganglionic axons then leave the celiac ganglion and travel with the common hepatic artery to the gastroduodenal artery. The postganglionic fibers then follow the branches of the gastroduodenal artery into the duodenum where they terminate.

V. GVA neurons and fibers associated with the sympathetics
The general first rule is that wherever there is a sympathetic GVE fiber, there will also be a GVA fiber. (One exception is the grey ramus communicans which contains the sympathetic postganglionic axons that re-enter the dorsal and ventral primary rami.) All viscera, including sweat glands and blood vessels are innervated by sensory nerves. The reason for this is that your nervous system needs to know if food is present in the gastrointestinal tract, how dilated blood vessels are, how well viscera are being oxygenated, etc. GVA fibers inform the nervous system of these conditions. GVA fibers make their way back through spinal nerves, splanchnic nerves and cardiac nerves to the sympathetic chain ganglia. Remember that while GVA fibers travel through ganglia to reach the spinal cord, they do not synapse in chain or dorsal root ganglia. The GVA fibers then travel through sympathetic chain ganglia, through the white rami and into the dorsal roots. As a rule, all GVA cell bodies from fibers that have passed through the sympathetic chain are located in the dorsal root ganglia. From the cell body, the central process travels through the dorsal root to enter the dorsal horn of the spinal cord. Remember that for sensory information to get from an organ to the spinal cord, only one GVA neuron is required.

VI. Organization of the parasympathetic division of the ANS

A. The parasympathetics are also referred to as the craniosacral division of the ANS. The name indicates the position of the parasympathetic preganglionic cell bodies. Some preganglionic neurons are located in the brainstem associated with cranial nerves III, VII, IX and X, while others are in the sacral parasympathetic nucleus in the S_2 to S_4 spinal cord segments. (See section on CRANIAL NERVES for a discussion of the parasympathetic components of cranial nerves III, VII and IX.) Remember, that there are no parasympathetic fibers in the body walls or limbs.

B. The vagus - GVE
The vagus nerve innervates the heart, lungs, abdominal organs down to the left colic (splenic) flexure and portions of the genitourinary system. Vagal parasympathetic preganglionic GVE cell bodies are located bilaterally in the medulla of the brainstem. Axons from these neurons exit the base of the skull as the right and left vagus nerves (vagi). These preganglionic fibers travel through the neck into the thorax where approximately 3 twigs from each vagal trunk head for the deep and superficial cardiac ganglia where they will synapse on postganglionic parasympathetic cell bodies innervating the heart and lungs. The vagal trunks will continue onto the surface of the esophagus where they split into a plexus. At the caudal end of the esophagus, the vagal trunks reform into an anterior trunk (which is derived from the left vagus) and a posterior vagal trunk (derived from the right vagus). The anterior vagal trunk divides into several gastric and hepatic branches which enter their respective organs. Once in the organ, the preganglionic axon terminates on a parasympathetic postganglionic neuron located within the muscular wall of the organ. The posterior vagal trunk pierces the diaphragm and largely enters the celiac ganglion. Remember that the celiac ganglion contains postganglionic sympathetic cell bodies. For ease of understanding, the preaortic ganglia can be considered as containing no parasympathetic postganglionic perikarya. The preganglionic parasympathetic fibers continue on through the celiac ganglia, without synapsing, and catch a ride on the artery of the organ to which they are supplying, i.e., preganglionic parasympathetic fibers to the spleen will form a plexus around the splenic artery. Once the preganglionic fiber reaches its organ, it will synapse on a postganglionic parasympathetic cell body which will lie on or within the organ itself. The postganglionic parasympathetic axon will then trail off and innervate the musculature of the organ.

C. GVA neurons and fibers associated with the vagus
As in the case with the sympathetics, GVA fibers also travel with the vagus nerves. GVA fibers associated with parasympathetic GVE nerves to the heart and lungs are involved with reflexogenic regulation of heart rate and bronchodilation. In the abdomen, parasympathetic GVA fibers inform the CNS of the presence of food in the bowel, intestinal distension, etc. This contributes to reflex mechanisms that propel the food bolus into the next bowel segment. Remember that only one GVA neuron is required to get an impulse from the organ back to the CNS. The processes of vagal afferents travel back up through the preaortic ganglia, through the thoracic vagal trunks into the vagus nerves in the neck. Just before the vagus enters the cranium, there is a large swelling on the nerve. Within this swelling, termed the nodose ganglion, are ALL of the perikarya of GVA vagal afferents.

D. S_2 to S_4 spinal cord segments - GVE
Parasympathetic preganglionic cell bodies to the lower gastrointestinal tract and to pelvic organs are located in the sacral parasympathetic nucleus between the S_2 and S_4 spinal cord segments. Preganglionic axons from these neurons project through the ventral root into the

spinal nerve. Remember that in the sacrum the spinal nerve branches before exiting the vertebral canal such that the dorsal primary rami exit the dorsal foramina of the sacrum while the ventral primary rami exit the ventral sacral foramina. As the spinal nerve branches into dorsal and ventral primary rami, the preganglionic parasympathetic fibers stay within the ventral primary ramus. Once out of the ventral sacral foramina, the preganglionic parasympathetic fibers branch off of the S_2 through S_4 ventral primary rami as the pelvic splanchnic nerves. Understand that these pelvic splanchnic nerves have absolutely nothing to do with the sympathetic chain.

Some axons in the pelvic splanchnic nerves synapse on postganglionic parasympathetic neurons located in the ganglionated inferior hypogastric plexus. The inferior hypogastric plexus is formed as the preaortic plexus extends caudally into the pelvis, it divides into right and left superior hypogastric plexi (also referred to as the hypogastric nerves). The caudal-most extend of each superior hypogastric nerve is termed the inferior hypogastric plexus.

Other axons enter the inferior hypogastric plexus and ascend to the inferior mesenteric artery. The fibers will then follow the branches of the inferior mesenteric artery out to parasympathetic postganglionic perikarya located the descending and sigmoid colons and the rectum.

E.   GVA neurons and fibers associated with the sacral parasympathetics
There are GVA fibers associated with the parasympathetics at S_2 to S_4. These fibers travel back from lower abdominal or pelvic viscera, through the hypogastric plexi, through the pelvic splanchnics, through the ventral primary rami, through the spinal nerve and into the dorsal root. The cell bodies of sacral GVA neurons are located in the S_2 through S_4 dorsal root ganglia. From the dorsal root ganglia, the central processes travel through the dorsal root into the dorsal horn of the spinal cord.

MICROANATOMY

CYTOLOGY

I.   Features common to most cells.

   A.   **Plasma membrane** (unit membrane). About 8 nanometers (nm) thick, lipid bilayer
        (phospholipids and cholesterol) with integral and transmembrane proteins, and
        carbohydrates. Carbohydrate units attached to proteins form the glycocalyx. Contains
        receptors, enzymes for transport, and forms a semipermeable barrier delimiting the
        intracellular environment (or nuclear or organellar environment) from an outside
        environment. Is specialized at many intercellular junctions (synapse, gap junction, tight
        junction, etc.).

   B.   **Nucleus**. Contains chromatin (DNA) and synthesizes messenger RNA (mRNA) and transfer
        RNA (tRNA).
        1.   Chromatin.
             a)   **Euchromatin**. Uncoiled chromatin active in DNA transcription, stains poorly with
                  histologic stains.
             b)   **Heterochromatin**. Condensed chromatin inactive in DNA transcription, well-
                  stained with basophilic dyes, electron-dense in electron micrographs, often
                  adjacent to the nuclear envelope.
        2.   **Nucleolus**. Synthesizes ribosomal RNA (rRNA), stains well with basophilic dyes.
             a)   Nucleonema. Pars granulosa (granular RNA that may represent maturing
                  ribosomes) and pars fibrosa (filamentous RNA which may represent primary
                  transcripts of rRNA genes).
             b)   Pars amorpha: nucleolar organizer DNA (codes for rRNA).
        3.   **Nuclear evelope**. Double unit membrane, pores for transit of m-, t-, and rRNA and
             other materials between nuclear and cytoplasmic compartments; outer membrane
             studded with ribosomes.

II. Features that are variable depending on cell type and function.

   A.   A. Cytoplasmic organelles that are membrane constructed.
        1.   **Endoplasmic reticulum** (ER). Appears as stacks of membranes or networks of
             tubules containing an inner space or lumen = cisterna.
             a)   **Rough** (rER or granularER, gER). Ribosomes (15nm) are attached to membrane.
                  (1)   **Ribosomes**. Particles composed of rRNA and protein forming two subunits.
                        Ribosomes are attached to strands of mRNA forming polysomes. Polysomes
                        can be free in the cytoplasm or attached to rER. The mRNA codes for
                        protein synthesis and calls for appropriate amino acids which are coded and
                        transported by tRNA. Free polysomes produce proteins for intracellular use,
                        whereas polysomes attached to rER produce proteins for export and for
                        lysosomal enzymes.
                  (2)   Peptides assembled at rER are temporarily stored in cisternae of rER, then
                        moved by transfer vesicles to the Golgi complex for modification and
                        packaging. rER is prominent in protein secreting organs and cells (e.g.,
                        salivary glands, pancreas, certain endocrine glands, plasma cells and
                        neurons).

b) **Smooth** (sER or agranularER). Lacks ribosomes so membranes are smooth. Prominence and function varies with cell type. Involved in Ca^{++} regulation in muscle (sarcoplasmic reticulum), detoxification of drugs in liver hepatocytes, lipid metabolism and synthesis of steroid hormones in certain endocrine glands.

2. **Golgi complex.** Stacks of smooth membranous cisternae and vesicles. Receives transfer vesicles containing immature protein from rER. Packages protein for export into secretory (zymogen) granules. Also involved in membrane sorting and recycling and addition of carbohydrate units to proteins.

3. **Lysosomes.** Contain lytic enzymes for intracellular digestion (however, some lysosomal enzymes are secreted for extracellular use). Includes ribonucleases, deoxyribonucleases, acid phosphatase, proteases, glycosidases, esterases and these are active mainly at acidic pH. Lysosomes are produced by the Golgi complex.
   a) Two types of lysosomes.
      (1) Primary - have not engaged in hydrolytic activity
      (2) Secondary - are or have been active in enzymatic activity (digestion).
   b) Interactions of lysosomes with materials.
      (1) **Phagocytosis** - ingestion of particulate material and sequestration into a membrane-bound vesicle.
         (a) Heterophagocytosis - ingestion of material from outside the cell (an antigen) forming a heterophagosome.
         (b) Autophagocytosis - ingestion of material from inside the cell (e.g., a damaged mitochondrion) forming an autophagosome.
         (c) Crinophagocytosis - ingestion of excess secretory product within a gland cell.
      (2) Pinocytosis - (endocytosis) ingestion of fluid materials and sequestration into membrane-bound vesicles.
      (3) Primary lysosomes fuse with phagosomes, dump enzymes into the phagosome and the enzymatic digestive process begins - membrane-bound structure is now called a **secondary lysosome**. If the process of digestion is not complete, then the remaining membrane-bound material and structure = **residual body** (dense body). These may accumulate as lipofuscin granules in long-lived cells (e.g., neurons and cardiac muscle cells).

4. **Peroxisome** (microbody). Membrane-bound intracellular structures that resemble lysosomes.
   a) Have a granular matrix and contain enzymes that reduce oxygen to hydrogen peroxide (hydrogen peroxide is toxic).
   b) Active in oxidative functions; contain the enzyme catalase that breaks down hydrogen peroxide to oxygen and water.
   c) Are prominent in cells that produce steroid hormones.

5. **Mitochondria.** Involved in production of energy as ATP. Numerous in metabolically active cells (e.g., striated muscle cells, neurons, and hepatocytes).
   a) Structure. Double-membrane bound. Outer membrane is unremarkable, but the inner one is thrown into folds or shelves called cristae. Inner membrane bounds the mitochondrial matrix. Granules in the matrix bind ions (especially calcium). Matrix contains DNA and ribosomes; DNA is important for replication of the mitochondria and production of mitochondrial proteins. Proteins produced here are for mitochondrial use.

b) Inner membrane possessess elementary particles which contain the enzymes for oxidative phosphorylation and ATP synthesis, cytochrome enzymes for electron transport and succinate dehydrogenase (part of TCA, Krebs, cycle).

c) Matrix contains other enzymes for TCA cycle.

d) Relation of structure to function. Mitochondria are self-replicating (involves DNA synthesis and then division of the mitochondrion with repair of membranes), contain their own DNA and machinery for protein synthesis. Mitochondrial DNA codes for mitochondrial enzymatic proteins (not structural proteins), ribosomes and RNA. Main function is production of ATP for protein systhesis, for intracellular movements and for muscle contraction.

B. Other cytoplasmic organelles that are not membrane constructed.

1. **Centrioles**. Paired, located near the nucleus, constructed of nine sets of triplet microtubules surrounded by an amorphous material (pericentriolar satellite material - gives rise to new procentrioles and basal bodies of cilia). Centrioles organize the microtubules of the mitotic spindle, determine the polarity of cells, and produce basal bodies of cilia during ciliogenesis.

2. **Cytoskeletal components**.
   a) **Microtubules**. 25nm thick, indefinite length and consist of the protein tubulin. Function in intracellular transport and support; prominent in mitotic spindles, cilia and flagella.
   b) **Filaments**.
      (1) **Microfilaments**. 5-7nm thick, consist of the protein actin, participate in contraction in muscle and many non-muscle cells.
      (2) **Intermediate filaments**. 8-11nm thick, composed of various proteins, depending on cell type.
         (a) Glial fibrillary acidic protein - astrocytes in the CNS.
         (b) Vimentin - fibroblasts, chrondroblasts, macrophages and vascular smooth muscle cells.
         (c) Desmin (skeletin) - muscle cells (except vascular smooth muscle cells).
         (d) Cytokeratins (prekeratin, tonofilaments) - epithelial cells.
         (e) Neurofilaments - neurons.
      (3) **Thick filaments**. 15nm thick, consist of the protein myosin, participate in contraction in muscle cells and some non-muscle cells.
      (4) Filaments generally have a supportive function within the cell and stabilize cellular specializations (e.g., microvilli and specialized junctions).

3. **Inclusions**. Temporary or stored elements in the cytoplasm that may or may not be membrane-bound.
   a) Lipid - present as droplets either membrane-bound or without a membrane.
   b) Glycogen - polymers of glucose; present as beta-granules or alpha clusters.
   c) Lipofuscin - membrane-bound; part of lysosomal system.
   d) Pigment granules - membrane-bound, e.g., melanin.
   e) Secretory granules - stored synthetic products, usually membrane-bound, that are released on stimulation, e.g., hormones, neurotransmitters and digestive enzymes.

C. Specializations of the cell surface.

1. **Microvilli**. Finger-like projections at the cell surface. Covered by the plasma membrane and glycocalyx (carbohydrate units attached to the protein of the membrane - important in cell-cell recognition and protection of the membrane). Have a core of

actin filaments that promote small movements of the microvilli. Function: increase surface area for exchange purposes and for absorption, also aids phagocytosis. Prominent and regular at the apical surface of intestinal epithelial cells and cells forming the proximal convoluted tubules of the kidney. Irregular at the surface of many cells.

2. **Stereocilia**. Are very long microvilli. Prominent on the epithelial cells of the epididymus.
3. **Cilia**. Are longer than microvilli and have microtubules rather than filaments. Shaft projects from the apical surface of ciliated epithelial cells and contains a core of microtubules arranged as 9 peripheral doublets surrounding a central pair (9+2 arrangement = axoneme). Basal body exists in apical cytoplasm of cells and resembles a centriole. Rootlet anchors the cilium into the cytoplasm.
4. **Junctions**. Specializations associated with the membrane of adjacent cells.
    a) **Junctional complex**. Occurs between epithelial cells; usually at the apical surface adjacent to a lumen. Consists of 3 components:
        (1) Tight junction (zonula occludens). 20nm intercellular space is obliterated and outer leaflets of membranes of adjacent cells fuse. Forms a zone or band around the cells. Low electrial resistence here so it is a means of communication.
        (2) Intermediate junction (zonula adherens). 20nm intercellular space is resumed. Tonofilaments insert into dense material along the junction membrane for stabilization. Mechanical coupling between cells.
        (3) Desmosome (macula adherens). Tonofilaments insert into the submembrane dense material. Small projections of the tonofilaments (transmembrane linkers) enter the intercellular space to aid in mechanical coupling. May occur independently from 1) and 2) - especially prominent in the epidermis.
    b) **Hemidesmosome**. One-half of a desmosome occuring between epithelial cells and the underlying connective tissue. Mechanically couples the epithelium to the connective tissue; especially prominent coupling the epidermis to the dermis.
    c) **Gap junction** (nexus). Close association between membranes of adjacent cells. Intercellular space narrowed from 20 nm to a 2nm "gap". Transmembrane particles (connexons) project from the membrane of one cell to that of the other. Connexons have pores that allow transfer of small particles (e.g., ions and cAMP) between cells to aid in communication. Common in excitable tissues (smooth muscle; part of intercalated discs of cardiac muscle; also found in nervous tissue).
5. **Vesicular system**. Invaginations of the cell membrane that aid in transport in/out of the cell.
    a) Pinocytotic (endocytotic) vesicles. Invaginations of the surface membrane that pinch off as 50nm vesicles and carry fluid materials into the cytoplasm. Prominent in endothelial cells of capillaries. Some are contain a coating of clathrin that confers some specificity in their transport=**coated vesicles**.
    b) Exocytotic vesicles. Appear like pinocytotic vesicles, but extrude material from the cell. They fuse with the surface membrane, become confluent with it and their lumen opens to the exterior to release contents. Also prominent in capillaries as well as kidney tubules.
    c) Phagocytotic vesicles. Infoldings of the surface membrane like pinocytotic vesicles, but are often larger and take-up particulate matter, e.g., cellular debris, bacteria, dust and smoke particles. Prominent in macrophages.

6. **Basal membrane infoldings**. Deep parallel invaginations of the basal membrane of certain epithelial cells (distal and proximal tubule cells of the kidney and duct cells of salivary glands). They are lined by mitochondria and serve to increase the surface area available for the active transport of fluids. Mitochondria provide energy for this process.

D. **Cell cycle**. Renewable cells undergo alternating periods of division (mitosis - visible cell division to produce daughter cells) and synthetic phases (synthesis of material, duplication of DNA, in prepartation for mitosis). DNA replicates during a phase called interphase (phase between divisions). The alternation between interphase and mitosis = cell cycle.
   1. **Interphase** has 3 phases. G_1(gap phase 1) - pre-DNA synthesis, but time for RNA and protein synthesis in preparation for mitosis; S-phase - (DNA synthesis phase); G_2 (gap phase 2) - post-DNA duplication and period of accumulation of energy for mitosis and synthesis of tubulin for microtubules of the mitotic spindle.
   2. **Mitosis** = M-phase. Actual division occurs.
   3. Some cells do not undergo mitosis (neurons and cardiac muscle cells) and are suspended in a phase called G_0 rather than go through the cycle.

E. **Mitosis**: M-phase. Nuclear membrane disintegrates, chromosomes condense (= **prophase**) and align on the equatorial plane (= **metaphase**), chromosomes move to the opposite poles (= **anaphase**), and cytokinesis (division of the cytoplasm) occurs with reassembly of the nuclear membrane (= **telophase**).

EPITHELIUM

I. There are two broad classifications of epithelium, surfacing (membranous) and glandular.

A. **Surfacing epithelium** covers the exterior of the body and lines all of the spaces inside the body. **NOTE**: Any material that enters or exits the tissues of the body must cross an epithelium. Two important characteristics of epithelium are that it is avascular and it has a high regenerative capacity. All epithelia are constantly being renewed, repaired and replaced; some at a faster rate than others. The entire epithelial lining of the GI tract is replaced approximately every 4 days. (Consideration of various types of surfacing epithelia will be addressed with specific organs and systems).

B. **Glandular epithelium**. Includes exocrine and endocrine glands.
   1. **Exocrine** gland cells are organized into groups = acini (alveoli) or tubules. These groups of cells secrete into a duct system. Duct cells may add to/modify the final secretion.
      a) Types of gland cells - serous and mucous.
         (1) Serous cells. Have a round prominent nucleus, abundant rER and Golgi complex, and secretion granules (zymogen granules) at their apical surfaces. Secretion is usually thin and contains enzymes. Example, parotid gland and pancreas.
         (2) Mucous gland cells. Have a flat, basally located nucleus, basal rER, prominent Golgi complex, and a light staining cytoplasm containing mucous secretory droplets (largely consisting of glycoprotein). Secretion is usually thick. Example, sublingual salivary gland.
         (3) Myoepithelial cells are located between the secretory cell and its basement membrane. Actin-containing processes embrace the secretory cell; upon contraction the myoepithelial cell helps express the secretory product.

162

b) Modes of secretion.
   (1) Holocrine. Whole cell is secreted; e.g., sebaceous gland.
   (2) Merocrine (eccrine). Secretory granules released; e.g., pancreas.
   (3) Apocrine. Some apical cytoplasm lost with the secretory product; e.g., mammary gland.
2. Endocrine gland cells - no ducts, secretion is released into capillaries and is distributed by the vascular system. **NOTE**: high vascularity, fenestrated capillaries or sinusoids present. Some cells store secretory product, some synthesize and release product on demand.

II. Epithelial membranes. Wet (lubricated) membranes; mucous and serous.

A. **Mucous membrane**. Single layer of epithelial cells plus the underlying thin layer of loose connective tissue (lamina propria of the GI tract). Lubricated by mucus-secreting epithelial cells (goblet cells). Lining of GI tract is a mucous membrane.

B. **Serous membrane** (serosa). Single layer of squamous epithelial cells plus a thin underlying layer of loose connective tissue. Lubricated by a transudate of fluid from capillaries in the connective tissue. Examples, lining (visceral and parietal layers) of pleural, pericardial and peritoneal cavities. This simple squamous wet epithelium is also called **mesothelium**. Allows frictionless movement of one organ on another. **NOTE**: can become inflamed and very painful.

III. Epithelial-connective tissue interface: **basement membrane**. Two components to the basement membrane; **basal lamina** and **reticular lamina**.

A. **Basal lamina**: produced by the epithelial cells; comprised of type IV collagen, laminin and heparan sulfate. Two parts: lamina lucida (nearest to epithelial cells) and lamina densa. Function: selective filter for substances passing to epithelial cells and provides a scaffolding for organization of regenerating epithelial cells.

B. **Reticular lamina**: produced by fibroblasts of underlying connective tissue. Contains type III collagen = reticular fibers. Helps adhere underlying connective tissue to basal lamina and thus epithelium.

CONNECTIVE TISSUE PROPER

I. Types and functions are widespread and diverse.

A. Subepithelial loose connective tissue (lamina propria) - common site for immune reactions.

B. Submucosa of hollow organs is moderately dense and permits movement of the mucosa.

C. Reticular connective tissue provides a loose stromal framework for lymphoid organs, liver and bone marrow.

D. Dense regular connective tissuse forms tendons, ligaments and fascia; provides considerable support or leverage.

E. Adipose tissue is found in the hypodermis and is associated with numerous organs where it provides functional padding and serves as a nutritional store.

II. Constituents: cells and extracellular matrix.

A. Cells; those resident in connective tissue.
1. **Fibroblasts** (fibrocytes). Resident cells that produce and maintain collagen, elastic and reticular fibers. Play an important role in wound healing and scar formation.
2. **Mast cells**. Common in loose connective tissues near mucous membranes and often located adjacent to blood vessels. Have many cytoplasmic granules. Produce heparin (an anticoagulant), histamine (stimulates leakage from vessels and smooth muscle contraction) and eosinophil chemotactic factor (attracts eosinophils). These factors have roles in inflammatory reactions.
3. **Macrophages** (histiocytes). Fixed macrophages in loose connective tissues. Capable of phagocytosis and pinocytosis. Have many lysosomes. Play a key role in immune reactions and presentation of antigen-related materials to lymphocytes.
4. **Adipocytes**. Yellow fat, unilocular. Store triglycerides. Function in support, energy stores, and insulation. Quantities vary in different locations.

B. Cells; those that are immigrant (are transitory and migrate from the vascular system - usually in response to injury or inflammation in tissues).
1. **Macrophages**. Derived from bone-marrow-stem cells or monocytes.
2. **Granulocytes** (neutrophils, eosinophils and basophils), **agranulocytes** (monocytes and lymphocytes). Derived from circulating blood by diapedisis (movement of cells through the wall of postcapillary venules). Function in immune reactions to combat microorganisms or antigens that generate inflammation.
   a) Neutrophils. Contain specific neutrophilic granules; phagocytose bacteria; increased numbers are seen during acute infections.
   b) Eosinophils. Contain specific eosinophilic granules; phagocytose foreign particles like bacteria; increase in numbers during parasitic infections and allergic reactions.
   c) Basophils. Similar structure and function to mast cells. Increase in numbers during inflammatory reactions.
   d) Monocytes. Become macrophages.
   e) Lymphocytes. React to antigens and initiate immune responses.
3. **Plasma cells**. Derived from small B-lymphocytes. Produce antibodies in response to antigenic stimulation. Often found close to epithelial mucous membranes (especially the GI tract).

C. Extracellular matrix: fibers and ground substance.
1. Fibers
   a) **Collagen fibers** (collagen type I). Has great tensil strength and resists stretching. Procollagen is secreted by fibroblasts and is converted to tropocollagen. Microfibrils (consisting of tropocollagen molecules) assemble into collagen fibrils and these are assembled into collagen fibers. 13 different types of collagen have been described and are peculiar to various tissues and locations.
      (1) Type I. Found in the dermis, tendon, bone and fibrocartilage.
      (2) Type II. Found in hyaline and elastic cartilage.
      (3) Type III. Found in reticular fibers of liver, lymphoid organs and kidney.
      (4) Type IV. Found in the basal lamina.
      Collagen is degraded by the enzyme collagenase which can be secreted by several cell types including fibroblasts, neutrophils and macrophages.
      Macrophages may secrete collagenase to help them move through connective

164

tissues. Collagen turnover is usually slow, but increases during inflammation and tissue repair (wound healing).

- b) **Reticular fibers** (collagen type III). Thin and delicate fibers produced by fibroblasts and reticular cells. Form loose supporting framework (stroma) for loosely organized organs that have a great deal of cellular traffic, such as lymphoid organs.
- c) **Elastic fibers.** Produced by fibroblasts, smooth muscle cells and chondrocytes. Fibers distend when stretched and can recover their original dimension. Consist of microfibrils and an amorphous component (elastin). Seen in the walls of blood vessels where they allow the vessel wall to expand with changes in blood pressure. In large vessels they form thick elastic laminae.

2. Ground substance. Amorphous intercellular substance.

- a) **Proteoglycans, glycosaminoglycans (GAGs)** and **glycoproteins**. Proteoglycan consists of a core protein to which GAGs are attached. GAGS include hyaluronic acid, chondroitin sulfate, dermatan sulfate, heparan sulfate and keratan sulfate.
- b) Proteoglycan function. These molecules are often bound to extracellular fibers to provide support. They attract and hold tissue fluid to provide a watery mesh that allows the movement of cells and molecules (e.g., water, metabolites and drugs) through the fibrous framework. They obstruct the movement of large molecules but permit movement of small molecules. Mineralization can occur in the ground substance, e.g., in bone.
- c) Fibronectin. Is a structural glycoprotein of connective tissue produced by fibroblasts. Helps bind the fibroblast to collagen to anchor the cell and provide attachments for movements.
- d) Tissue fluid of connective tissue is derived from the blood. This material is forced out of the capillary while cells and large molecules are retained in the capillary. This provides an avenue for metabolites to exit the vascular system to the connective tissues.

CARTILAGE

I. Several types and functions: hyaline, elastic and fibrocartilage generally provide mechanical support.

II. Constituents: cells and extracellular matrix.

- A. **Cells**. Chondrocytes, chondroblasts and fibroblasts (the latter two cell types are mainly in the perichondrium).
  1. Chondrocytes are encased in extracellular matrix (fibers and ground substance) and occupy a space called lacunae.

- B. **Extracellular matrix**. Fibers and ground substance.
  1. **Fibers**. Mainly collagen (type II) and elastic, some reticular. Provide support and mechanical stability.
  2. **Ground substance**. Water and cartilage proteoglycans (GAGs = hyaluronic acid, chondroitin sulfate and keratan sulfate, attached to a core protein) and glycoprotein (chondronectin). Binds water and provides support by resisting deformation and compression.

        a)   GAGs form matrix. Highest concentration of GAGs is immediately around chondrocyte = capsule or territorial matrix. GAGs in between lacunae = interterritorial matrix.

        b)   Chondronectin. Attach type II collagen to chrondrocyte.

III.   Perichondrium. Special layer at the surface of cartilage. Two layers: outer fibrous layer = dense connective tissue and is protective; inner more cellular layer = chondrogenic (can give rise to new chondroblasts). **NOTE**: articular cartilage lacks a perichondrium.

IV.   Types of cartilage. Vary in content of fibers and constitution of ground substance.

     A.   **Hyaline**. Glassy matrix. Location: tracheal, costal (rib), and bronchial cartilages, and articular surfaces. Model for endochondral bone development. Articular cartilage = hyaline cartilage but lacks a perichondrium.

         1.   Constituents. Cells, fibers and ground substance. Cells = chondrocytes in lacunae. **Fibers** = mainly collagen type II. **Ground substance** = cartilage proteoglycan consisting of a core protein with attached GAGs (mainly hyaluronic acid, chondroitin sulfate and keratan sulfate). Content of sulfate groups with GAGs results in light bluish staining with H&E and metachromasia with toluidine blue.

     B.   **Elastic**. Appears similar to hyaline but matrix is more fibrillar. Location: pinna and external ear canal, epiglottis, and some laryngeal cartilage.

         1.   Constituents. **Cells** = chondrocytes in lacunae. **Fibers** = many elastic fibers but also type II collagen. **Ground substance** = like that of hyaline cartilage. Provides elasticity for movement.

     C.   **Fibrocartilage**. Resembles dense connective tissue, but has chondrocytes. Location: intervertebral discs, ligamentous attachments and pubic symphysis.

         1.   Constituents. **Cells** = chondrocytes in lacunae, but is less cellular than elastic or hyaline cartilage. **Fibers** = mainly collagen type I, but also type II collagen. **Ground substance** = fewer GAGs than elastic or hyaline cartilage, thus appears less basophilic.

V.   Growth of cartilage

     A.   **Interstitial**. Mitotic division of chondrocytes within lacunae. Daughter cells form nests or isogenous groups that secrete, and are surrounded by, the territorial matrix. Occurs developmentally and at the growth plate of bone for elongation.

     B.   **Appositional**. Mitotic division of chondrocytes of the inner layer of perichondrium (chrondrogenic layer). Daughter cells secrete territorial matrix to increase the girth of the cartilage.

BONE

Bone is organized and viewed as a tissue (connective tissue) and as an organ (for support and as a Ca^{++} store).

I.   Bone as a tissue.

     A.   Constituents: cells and extracellular matrix.

         1.   Cells: osteoblasts, osteocytes and osteoclasts.

         2.   Extracellular matrix: organic and inorganic components.

a) Organic constituents are fibers and ground substance. Fibers= collagen type I. Ground substance = proteoglycans, GAGs and glycoproteins.

b) Inorganic constituents are calcium phosphate organized as **hydroxyapatite crystals**, but also calcium carbonate and magnesium, citrate, sodium, potasium ions. Hydroxyapatite crystals organize along collagen fibers and are surrounded by ground substance to give bone its rigidity and strength.

B. **Osteoid**. Type I collagen and ground substance secreted by osteoblasts. This becomes mineralized around the osteoblast and the cell is now called an osteocyte. The osteocyte exists within a space of the mineralized bone = a **lacuna**. New osteoid may be secreted by osteoblasts to repeat the process and form layers of mineralized osseous tissue called **lamellae**.

C. Microscopic organization. Osseous tissue is organized in layers or lamellae. Lamellae are highly organized in compact (cortical) bone, but less organized in spongy (cancellous) bone.
1. **Compact bone.**
    a) **Periosteum**. Connective tissue that covers the external surface of compact bone. Two layers:
        (1) Inner layer is cellular and contains osteogenic cells.
        (2) Outer layer is fibrous. Collagen fibers here enter the bone matrix as Sharpey's fibers and hold periosteum to bone.
    b) **Endosteum**. Fine connective tissue, osteoblasts and osteoclasts that line all internal surfaces of both compact and spongy bone.
    c) Cells.
        (1) **Osteoblasts** at endosteal and periosteal surfaces secrete osteoid. Osteoid is laid down in lamellae, the matrix of osteoid becomes calcified. Cells trapped in calcified matrix = osteocytes.
        (2) **Osteocytes** are trapped in spaces of calcified matrix = lacunae. Osteocytes of adjacent lamellae are connected and communicate via their processes which are housed in mini-canals = **canaliculi**. Osteocyte processes connect and communicate via **gap junctions** (nexuses).
        (3) **Osteoclasts** are multinucleated cells found in marrow cavity, Haversian canal, endosteum and periosteum. Are crucial for remodelling, bone resorption and Ca^{++} mobilization. They are involved in bone matrix phagocytosis - secrete acid hydrolases to facilitate bone remodelling and Ca^{++} mobilization. Their activity is hormone sensitive: **parathyroid hormone** (parathormone) increases activity, whereas **calcitonin** decreases their activity.
    d) Lamellae. Concentric layers of calcified bone matrix organized around a neurovascular bundle (artery, vein and nerve) and endosteum. Neurovascular bundle and endosteum are in a canal = Haversian canal.
        (1) **Haversian system (osteon)**. Haversian canal + several lamellae. Runs longitudinal in long bone.
        (2) **Volkmann's canal**. Canal perpendicular to and connecting Haversian canals. House nutrient vessels coming from the periosteum.
        (3) Inner circumferential lamellae. Lamellae that arise from the endosteum and lay next to the marrow cavity.
        (4) Outer circumferential lamellae. Lamellae that arise from the periosteum and are at the periphery of the bone.

     (5) Interstitial lamellae. Lamellae that are between osteons and result from incomplete remodelling of older osteons.
  2. **Spongy bone.** Present as spicules in the core of long bones and is surrounded by compact bone. It lacks a periosteum but is surrounded by endosteum. Osteoid is laid down as lamellae, but osteons are not present.

II. Bone as an organ.
 A. Macroscopic structure.
  1. Types: long bones (e.g., femur, humerus), short bones (e.g., small bones of the hand and foot), and flat bones (e.g., calvarium, part of the mandible).
  2. **Long bone.** Consists of three named components.
   a) <u>Diaphysis</u>. Shaft or middle part, consists of compact bone and houses bone marrow cavity.
   b) <u>Metaphysis</u>. Expanded area at end of diaphysis, adjacent to growth plate or epiphyseal plate. Contains spongy bone.
   c) <u>Epiphysis</u>. End of long bone that articulates with another bone in a joint. Contains spongy bone. Covered by special hyaline cartilage = articular cartilage.
  3. **Short bone.** Consists of compact bone surrounding spongy bone.
  4. **Flat bones.** Two plates of compact bone containing a layer of spongy bone (diplöe).

 B. Bone formation, growth and remodelling.
  1. Osteogenesis. Bone forms by **intramembranous ossification** (flat bones and spicules arise from condensations of mesenchyme) and **endochondral ossification** (long and short bones arise from replacement of a hyaline cartilage model).
  2. **Intramembranous bone formation (ossification).**
   a) A condensation of mesenchymal connective tissue gives rise to fibroblast-like cells which form osteoblasts. Osteoblasts lay down osteoid within the mesenchymal connective tissue. Mineralization of osteoid occurs by deposition of calcium and phosphate salts on collagen.
   b) Osteoblasts that become trapped in their matrix are called osteocytes and possess processes within canaliculi which are connected to adjacent osteocytes by gap junctions. New layers of osteoid are laid down in a lamellar pattern.
   c) Few lamellae form and may give rise to spicules (of spongy bone). Matrix is maintained by osteocytes in the lamellae and osteoblasts at the spicule surface. Spicules are remodelled by osteoclasts at the surface.
   d) Some blood vessels may become trapped in forming bone with lamellae surrounding them. This gives rise to Haversian canals and eventually Haversian systems. Eventually this intramembranously formed spongy bone can be converted to compact bone (e.g., calvarium).
  3. **Endochrondral bone formation (ossification).**
   a) Characterized by mesenchyme giving rise to a hyaline cartilage model in areas where a bone is to exist. Cartilage model can grow by **appositional** (in width) and **interstitial** (in length) mechanisms.
   b) Ossification begins in diaphysis. As cartilage matures, many chondrocytes hypertrophy and die. Cartilage matrix calcifies and chondrocytes surrounded by calcified matrix also die leaving spaces (old lacunar spaces). The perichondrium around the altered cartilage differentiates into periosteum. Osteogenic cells in periosteum lay down a periosteal collar of bone around the diaphysis.

c) Periosteum (**periosteal bud**) grows into cavities of the cartilage created by dying chondrocytes and by confluence of lacunae. Periosteal bud contains osteogenic cells and blood vessels. Osteoblasts of the bud become positioned on the calcified cartilage spicules and start laying down osteoid. Thus, bone formation (spongy bone) begins at the **primary center of ossification** (in diaphysis).

d) Continued death of cartilage cells results in enlargement of the ossification center. Osteoclastic activity on the bone spicules leads to erosion and formation of the marrow cavity. Elongation of the cartilage model continues by interstitial growth of chondroblasts distal to zone of calcified cartilage/bone deposition.

e) **Secondary centers of ossification** occur in each epiphysis of long bones shortly after birth. A similar process proceeds as described above for the diaphysis: chondrocytes hypertrophy and die, cartilage matrix calcifies and is resorbed forming cavities containing calcified cartilage spicules, capillaries and osteogenic cells invade the cavity, and osteoblasts lay down osteoid on the cartilage spicules. This ossification process spreads in all directions, but leaves intact the articular cartilage at one end and the **epiphyseal growth plate** at the other.

f) The bone model continues to increase in length by interstitial growth of cartilage at the epiphyseal plate. The cartilage produced here continually calcifies and is replaced by bone. That bone is continually remodelled by osteoclastic activity.

g) The phases of endochrondral bone formation can be visualized at the **epiphyseal growth plate**:
   (1) Resting or reserve zone.
   (2) Proliferative zone (interstitial growth).
   (3) Zone of maturation and hypertrophy.
   (4) Zone of calcification of the cartilage matrix and chondrocyte death.
   (5) Zone of ossification with deposition of bone on cartilaginous spicules.
   (6) Zone of resorption. Osteoclastic activity remodels newly formed bone

h) After puberty interstitial growth of the epiphyseal plate ceases and its location is indicated by the epiphyseal line.

i) Bone grows in diameter by appositional growth. Periosteum lays down bone at the diaphyseal surface, whereas bone is resorbed (by osteoclasts) from the inside of the bone. Thus, the marrow cavity is enlarged in the diaphysis.

C. Factors affecting bone growth.
   1. Gonadal hormones. Influence normal rates of skeletal maturation. **NOTE**: postmenopausal drop in estrogen leads to increased bone resorption = osteoporosis.
   2. Growth hormone. Promotes proliferation of chondrocytes at epiphyseal plate.
   3. Vitamins. Vitamin C is important for collagen synthesis; vitamin D, promotes Ca^{++} absorption in the intestine and promotes calcification and collagen synthesis.

PERIPHERAL BLOOD

I.   Blood is a liquid connective tissue that has numerous functions including: transport of O_2, CO_2, hormones and metabolites, is involved in acid-base balance, immunologic defense and temperature control. **NOTE**: blood is a window into the body because it is easily accessible and study of its chemistry and cellular content and morphology is a view of a patient's state of health.

II. Constituents: cells (formed elements) = 45% and extracellular matrix (plasma) = 55%.

   A.   Cells: erythrocytes (red blood cell, RBC), leukocytes (white blood cells, WBC) and platelets. Origin: bone marrow.
      1.   **Erythrocytes (RBCs).**
         a)   Biconcave disk in shape, increases surface to volume ratio to promote gas exchange. Abnormal shape = poikilocyte (poikilocytosis = sickle cell). Number = $5\text{-}10\text{x}10^6$ per mm^3 of blood (reduced number = anemia; increased number = polycythemia). Size = 7.5µm diameter normally; macrocytes > 9µm; microcytes < 6µm. Lifespan = 120 days. Does not contain any organelles, but does contain a cytoskeletal protein, spectrin, which helps maintain shape.
         b)   Major constituent is **hemoglobin** = a protein with high O_2 binding capacity. Hemoglobin + O_2 = oxyhemoglobin (red in color); oxyhemoglobin minus O_2 = reduced hemoglobin (blue in color). O_2 binds to the heme part of the hemoglobin molecule.
         c)   Reticulocyte. Immature RBC released from bone marrow, usually when there is great demand for new RBCs. May constitute 1% of circulating RBCs.
      2.   **Leukocytes** (WBCs): granulocytes and agranulocytes.
         a)   Granulocytes. Comrprise neutrophils, eosinophils and basophils. Contain specific granules (contents specific to each leukocyte) and non-specific granules (azurophilic granules = lysosomes).
         b)   Agranulocytes. Comprise lymphocytes and monocytes. Lack specific granules, but contain lysosomes.
         c)   **Neutrophils** (polymorphonuclear leukocytes, PMNs). Up to 60-70% of circulating WBCs.
            (1)   Round shape. Number = 4,500 per mm^3 of blood. Size = 12-15µm diameter. Lifespan = 12-14 hrs in blood and 1-4 days in connective tissue. Nucleus is multilobed (3-5 lobes in adult cells). Barr body or 'drumstick' is sex chromation evident associated with the nucleus of up to 3% of PMNs examined in females. PMNs function in the loose connective tissues and not in blood; exit vascular system by **diapedesis** from **postcapillary venules**.
            (2)   Major cytoplasmic constituent = membrane-bound granules.
               (a)   Non-specific granules (azurophilic granules = lysosomes). Form 20% of the granules and contain hydrolytic enzymes.
               (b)   Specific granules. Comprise 80% of the granules and contain bacteriocidal agents and alkaline phosphatase.
            (3)   Function. Phagocytosis and destruction of bacteria. Specific granules fuse with formed phagosome and contents of specific granules (especially lysozyme) disrupt bacterial cell wall. Then azurophilic granules help digest contents of the phagosome. Bacterial infections will lead to increased

numbers of PMNs and increased numbers of immature ('band') cells in the circulation.

d) **Eosinophils**. 2-4% of circulating WBCs.
- (1) Round shape. Number = 200 per mm^3. Size = 12-15μm diameter. Nucleus has 2-3 lobes. Lifespan of a few hours in blood before they exit to connective tissue. These are mobile, phagocytic cells.
- (2) Major cytoplasmic constituent = membrane-bound granules. Specific granules stain red-orange with eosin. Granules have a prominent cytstalline core containing myelin basic protein (has an antiparasitic function). Granules are lysosomes rich in peroxidase.
- (3) Function. Eosinophils kill parasitic larvae and increase in number during parasitic infections. Lysosomal enzymes help destroy dead parasites. They also phagocytose antigen-antibody complexes. Thus, they are increased in number during such allergic reactions as hay fever.

e) **Basophils**. 1% of circulating WBCs.
- (1) Round cells. Number = 5 per mm^3 of blood. Size = 12-15μm diameter. Nucleus has 2-3 lobes. Basophils function in connective tissue.
- (2) Major cytoplasmic constituent = membrane-bound granules.
  - (a) Specific granules stain deep purple-magenta. They contain a crystalline substructure which suggests they are lysosomes. Granules contain heparin (anticoagulant), histamine (fast vasodilator) and slow reacting substance (slow vasodilator).
- (3) Function. Basophils function similar to mast cells in inflammatory reactions; participate in inducing inflammatory reactions. They can bind immunoglobulin E (IgE) that is produced by plasma cells in response to an antigen, but have no response until they see that antigen again - then they react quickly and the released histamine can cause severe reactions (i.e., as in bronchial asthma attack or anaphylactic response after a bee sting). These are examples of immediate hypersensitivity.

f) **Lymphocytes**. 20-30% of circulating WBCs.
- (1) Round cells. Number = 2500 per mm^3 of blood. Size = 6-8μm in diameter for small lymphocytes in circulation. Small spherical basophilic nucleus. Thin rim of cytoplasm containing few organelles. Some azurophilic granules (lysosomes). Function in connective tissue. Two types of lymphocytes (functionally).
  - (a) B-lymphocyte. Develop in the bone marrow. They give rise to plasma cells and participate in the humoral immune response, i.e., production of circulating antibodies.
  - (b) T-lymphocyte. Become immunocompetent in the thymus gland and participate in the cell-mediated immune response, e.g., graft rejection.

g) **Monocytes**. 3-8% of circulating WBCs.
- (1) Round-oval cells. Number = 300 per mm^3 of blood. Largest of the WBCs, size = 12-18μm. Large, light-staining, slightly indented nucleus with a delicate network of chromatin evident. Cytoplasm is abundant and contains azurophilic granules (lysosomes). Monocytes exit the vasculature to function in connective tissue.
- (2) Function. Monocytes are the precursors to tissue **macrophages** (histiocytes) and other macrophages that leave tissues, e.g., alveolar macrophages lining air spaces of the lung. As macrophages they can present antigens to

lymphocytes and participate in immune reactions. Lifespan of 1-2 days in circulation and up to 60 days as macrophages.

3. **Platelets** (thrombocytes). Are anucleate parts of cells circulating in peripheral blood.

   a) Oval to rounded particles. 200,000-400,000 per mm^3 of blood. Size = 2-4µm in diameter. Derived from megakaryocytes in bone marrow. Cytoplasm has a system of microfilaments (actin) and microtubules to help maintain shape. Also many granules in the cytoplasm.

   b) Granules = alpha granules containing <u>clotting factors</u> and dense-core granules containing <u>serotonin</u>.

   c) Function. Play a role in blood clotting. Platelets aggregate at a defect in the endothelium of injured blood vessels. Platelets release serotonin to cause constriction to reduce blood flow and clotting factors (thromboplastin). Thromboplastin is involved in the conversion of prothrombin to thrombin and the latter converts fibrinogen to fibrin. Fibrin forms threads to trap other platelets and blood cells to form a clot = thrombus. Life span of platelets is 7-10 days.

4. **Hematocrit**: clinical estimate of the two major components of blood, the cells vs. plasma expressed as an estimation of the volume of packed RBCs per unit of blood volume. Centrifuge blood and RBCs pack to the bottom. Normally about 45% of the sample is RBCs and 55% is plasma. A **buffy coat** of WBCs is evident between the RBCs and plasma. **NOTE**: hemocrits vary with gender and age.

5. **Differential leukocyte count**. Count 100 consecutive leukocytes and classify them as to cell type. Percentage is derived directly from the number and should be about 70% neutrophils, 3% eosinophils, 1% basophils, 20% lymphocytes and 6% monocytes.

B. Extracellular matrix. Comprised of fibers (potential) and ground substance.

   1. Fibers (potential). Fibrinogen can become fibers upon induction of clotting.

   2. Ground substance. Comprised of plasma (fluids and proteins). Proteins include albumin, gamma globulins and fibrinogen. Also, organic compounds such as amino acids, vitamins, hormones and lipoproteins are found in plasma as well as inorganic salts. However, about 90% of plasma is H_2O.

   3. **Serum** is the fluid that remains after a clot is formed. Serum is essentially plasma minus fibrinogen and cells.

HEMATOPOIESIS

I. Origin and development of blood cells.

A. Sites: extraembryonic and intraembryonic.

   1. **Extraembryonic**. Occurs in blood islands in the yolk sac during second week of gestation.

   2. **Intraembryonic**. Occurs in liver and spleen from second to sixth months <u>and</u> in bone marrow from mid-gestation into adulthood.

B. Bone marrow. **Red bone marrow** is hematopoietic (color due to erythroid elements and hemoglobin) and is present in flat bones (e.g., calvaria and sternum) and long bones. **Yellow bone marrow** is not hematopoietic in adult (color due to fat content).

II. **Red bone marrow tissue** is of mesenchymal origin.

A. Structure: stroma is a 3-D meshwork of reticular cells and fibers (collagen type III). Some reticular cells are phagocytic. Fat cells are present as well. Hematopoietic cells occupy meshes of stroma. Vascular channels (sinusoids lined by endothelial cells) are present throughout the stroma and among the hematopoietic cells. As the blood cells develop, they migrate across the sinusoidal endothelium to enter the circulation.

B. Megakaryocytes are present near the sinusoidal endothelium. Platelets fragment from megakaryocytes and enter the sinusoids.

III. Hematopoiesis in red marrow.

A. Origin of cells. The theory that all blood cells develop from a single pluripotent stem cell = **monophyletic theory**. The stem cell = a **colony forming unit** (CFU).

B. The CFU gives rise to cells that respond to a **stimulating agent or poietin** (erythropoietin, granulopoietin or thrombopoietin).
   1. CFU cells responding to **erythropoietin** (produced in renal juxtaglomerular cells) give rise to erythroid cells.
      a) CFU cells give rise to **pronormoblasts**. Pronormoblasts have a prominent nucleus, mitochondria, ribosomes and a Golgi complex. Ribosome content increases, cytoplasm becomes more basophilic and cells are large (15μm diameter) = **basophilic normoblast**. Hemoglobin synthesis, begins in these cells.
      b) Hemoglobin accumulates in the cytoplasm and reduces the basophilia (cytoplasm becomes orange/grey), the cell nucleus condenses and this cell = **polychromatophilic normoblast**.
      c) As hemoglobin dominates the cytoplasm the basophilia is lost and the orange color of a mature RBC is evident, the nucleus is highly condensed (pyknotic) and cell becomes smaller (8-10μm diameter) = **orthochromatophilic normoblast**.
      d) The nucleus is shed along with any remaining mitiochondria; a few polyribsomes remain, but the cell can now enter the circulation = **reticulocyte**.
      e) Loss of all cellular organelles with maxiumum amount of hemoglobin and reduced size (7-8μm) = mature RBC.
      f) Time span from CFU to mature RBC = about 7 days.
   2. CFU cells responding to **granulopoietin** give rise to granulocytes.
      a) CFU gives rise to a **myeloblast**. Large cell (12-15μm diameter), no granules, 1-3 prominent nucleoli and lightly basophilic cytoplasm.
      b) **Promyelocyte**. Appearance of azure (non-specific) granules in cytoplasm, nucleoli prominent.
      c) **Myelocyte**. Specific granules appear and azure granules decrease. Nucleoli are not evident. Cells are identifiable as neutrophilic, eosinophilic or basophilic myelocytes.
      d) **Metamyelocyte**. Indentation of nucleus evident, but it is not condensed nor multilobed. Cells decreased in size (10-12μm). Specific granules are prominent.
      e) **Band cell** (stab or segmented cell). Is nearly mature and has a deeply indented or slightly lobulated nucleus. These can be released into the circulation (especially by bacterial infection). As nucleus becomes lobulated, then cells are mature neutrophils, eosinophils or basophils.
      f) Development time from CFU to mature granulocyte = about 12 days.

3. CFU cells (probably same as those producing granulocytes) give rise to monocytes.
    a) Developmental history in **monocytopoiesis** is different from granulocytopoiesis.
    b) Cells have monoblast, promonocyte and monocyte stages. No specific granules develope and nucleus does not condense and become lobulated.
    c) Monocytes leave the marrow via sinusoids, briefly enter the circulation, and exit to become tissue macrophages.
4. CFU cells (probably same as those producing granulocytes) give rise to lymphocytes.
    a) **Lymphocytopoiesis** occurs in bone marrow and in lymphoid organs and tissues. Large cells called lymphoblasts give rise to lymphocytes.
    b) Lymphocytopoiesis produces **T-lymphocytes** (which complete differentiation in the thymus and colonize lymphoid tissues and organs), **B-lymphocytes** (which are bone marrow derived, colonize lymphoid tissues and differentiate into plasma cells that produce antibodies), and **null lymphocytes** (which are cytotoxic cells that are not typical T- or B-lymphocytes).
5. CFU cells give rise to **megakaryocytes**.
    a) Have megakaryoblast and megakaryocyte stages.
    b) Megakaryocytes develop from megakaryoblasts. Megakaryoblasts undergo karyokinesis without cytokinesis. The megakaryocytes are large cells (100-150μm diameter) that have a large, multilobed, polyploid nucleus.
    c) Megakaryocyte cytoplasm developes a system of canaliculi so that as the cell cytoplasm extends across the wall of a sinusoid it fragments giving rise to **platelets** (thrombocytes).

IV. Summary. Of all the blood cells, RBCs, granulocytes, monocytes, platelets and B-lymphocytes are produced in the bone marrow and only T-lymphocytes are produced elsewhere, i.e., in the thymus.

IMMUNE SYSTEM (LYMPHATIC SYSTEM)

I. This is a system with multiple functions and diverse structure and organization.

   A. Function. Protect the individual from foreign material, e.g., antigens, bacteria, viruses and parts of "self" that have become abnormal.

   B. Cells: principally lymphocytes and macrophages.
       1. **Lymphocytes**. They must recognize "self" vs. "non-self"
           a) **B-cells**. Differentiate into plasma cells and secrete antibodies (immunoglobulins). They constitute about 20-25% of circulating lymphocytes.
           b) **T-cells**. Differentiate in thymus gland and are involved in cell-mediated responses (e.g., graft rejection). Form about 75-80% of circulating lymphocytes. Respond to antigens by differentiating into:
               (1) T-helper cells - stimulate B-lymphocyte differentiation.
               (2) T-supressor cells - supress B-lymphocyte differentiation.
               (3) T-killer cells - contact and kill foreign cells (grafts).
               (4) Natural killer (NK) cells - are larger than usual T-lymphocytes and directly attack foreign cells.
               (5) Null cells - difficult to identify these as T- or B-cells, but they can differentiate into killer cells.
       2. **Macrophages**. Participate in recognition and phagocytosis of antigens and cooperate with lymphocytes in immune responses.

II. Neutralization of antigens occurs by 2 main mechanisms.

    A. **Cell-mediated response**. T-cell directly attacks foreign material.

    B. **Humoral response**. B-cells differentiate into plasma cells and produce antibodies which combine with antigen. Some B-cells need to be "instructed" by T-cells or macrophages to differentiate into antibody producing plasma cells.

III. Components of lymphatic system: 4 broad elements.

    A. Aggregates of lymphoid tissue. Includes loose lymphatic tissue in loose connective tissue, tonsils containing lymph nodules, and aggregates of nodules as in Peyer's patches of the ileum.

    B. Lymphoid organs. Thymus, spleen and lymph nodes

    C. Lymph vessels. Capillaries and larger return vessels.

    D. Lymphocytes.

IV. Aggregates of lymphatic tissue.

    A. Loose lymphatic tissue. Is especially seen under the epithelia of the GI, respiratory and genitourinary systems (within the lamina propria).

        1. Its contents include all of the cellular elements, connective tissue fibers and ground substance seen in connective tissue proper. It also contains numerous efferent lymph capillaries to carry materials away from the area and toward regional lymph nodes.

        2. Of note is the abundance of **reticular cells and fibers** = stroma for all lymphatic organs. Important as well is the presence of small lymphocytes, macrophages, plasma cells, and mast cells.

        3. This is a primary site where immune reactions occur because antigens can easily cross the epithelia of organ systems exposed to the exterior. From here, responses are carried to regional lymph nodes via lymph capillaries and vessels.

    B. **Lymph nodule**. Non-encapsulated aggregate of primarily small lymphocytes; mainly B-cells. Often seen in loose lymphatic tissue or lymph organs

        1. Structure. May see a dark corona or marginal zone of tightly packed lymphocytes surrounding a lighter area (germinal center). Germinal centers appear after antigenic stimulation. Larger lighter staining "blast" forms of lymphocytes and plasma cells predominate in germinal centers.

        2. May be solitary or in aggregates. Aggregates are present in tonsils, appendix and **Peyer's patches of the ileum**, and cortex of lymph nodes.

    C. **Tonsils**. Aggregates of lymph nodules at the juncture of the oral and nasal passages.

        1. Three sets of tonsils form Waldeyer's ring of lymphoid tissue at entrance to GI and respiratory systems: <u>palatine tonsils</u> (located on the lateral side of the posterior oral cavity), <u>lingual tonsils</u> (located on the root of the tongue), and <u>pharyngeal tonsils (adenoids)</u> (located at the roof of the posterior nasopharynx).

        2. Structure: consist of invaginations of epithelium (crypts) lined by lymph nodules. Lymphocytes migrate across the epithelium and nearly obscure its identity.

V.   Lymphoid organs.

   A.   **Lymph nodes**. Designed to filter lymph derived from various connective tissues and spaces in the body.
   1.   Structure.
      a)   Supported by a collagenous framework consisting of a connective tissue capsule from which radiate collagenous trabeculae.
      b)   Parenchyma is supported by a stroma of **reticular cells and fibers** (this stroma is the same for all lymphoid tissues and organs as well as the liver).
      c)   Parenchyma consists of lymphocytes, macrophages, lymph nodules and lymph sinuses.
      d)   Afferent lymph vessels enter capsule. Efferent lymph vessels exit at hilus. Arteries, veins and nerves enter and exit the hilus.
   2.   Divisions: cortex and medulla.
      a)   **Cortex**. Packed with lymphocytes and lymph nodules (mainly B-lymphocytes). A system of lymph sinuses also are present in the cortex.
      b)   **Medulla**. Contains cords of cells (mainly B-lymphocytes). Also contains a vast system of lymph sinuses.
   3.   Circulation through the lymph node also provides a means of filtration of lymph derived from connective tissues of the body and other lymph nodes
      a)   **Afferent lymph vessels** enter capsule. They enter a system of sinuses for passage of lymph. Sinuses are lined by endothelial and reticular cells and macrophages that survey the passing lymph for foreign cells or antigens; such antigens will be phagocytosed.
      b)   Afferent vessel is continuous with a **subcapsular (marginal) sinus** which flows to a **trabecular (cortical) sinus** which becomes continuous with **medullary sinuses** (all sinuses are constructed the same and lined by the same cells as described above). Medullary sinuses drain into **efferent lymph vessels** that exit the node at the hilus.
   4.   Functional localization of lymphocytes.
      a)   **Paracortex** (deep cortex, thymus-dependent area) is populated by T-cells.
      b)   **Outer cortex** (containing lymph nodules) and **medulla** are populated by B-cells.
      c)   Most lymphocytes are not generated in the node, but enter the node by diapedesis through postcapillary venules.
   5.   Function. Lymph nodes filter lymph, trap and phagocytose antigens. Some lymphopoiesis occurs but this is minor. There is production of plasma cells from differentiation of B-cells, and thus, secretion of antibodies via plasma cells.

   B.   **Spleen**. Largest of lymphoid organs.
   1.   General function. Spleen is insinuated in the blood stream and is designed for filtering blood; it-clears particulate matter, dead cells, effete RBCs and is involved in immune reactions against blood-borne antigens.
   2.   Structure
      a)   Supported by a collagenous framework consisting of a thick collagenous connective tissue capsule and trabeculae. Trabeculae carry vessels and nerves into the depths of an organ. The parenchyma is supported by a delicate stroma of reticular cells and fibers.
      b)   Parenchyma. Lacks a cortex or medulla but, consists of a **red and white pulp**.
         (1)   **Red pulp**. Cords of cells = splenic or pulp cords or cords of Billroth; are organized around the venous system (venous sinusoids). Pulp cords are

very cellular and contain cells normally seen in loose lymphatic tissue and in circulating blood (i.e., reticular cells and fibers, plasma cells, macrophages, lymphocytes, monocytes, neutrophils and RBCs). Red pulp is designed to filter blood via its content of macrophages.

(2) **White pulp**. Splenic lymph nodules (with germinal centers) organized around the arterial system, specifically a central artery (arteriole). Mainly B-lymphoocytes in the nodules; can be involved in immune reactions. Arteries are surrounded by a sheath of lymphocytes, periarteriolar lymphatic sheath (PALS), mostly T-cells.

(a) At the edge of splenic nodules is a **marginal zone**. This is an interface between the red and white pulps. This is a site where T-cells derived from the blood encounter B-cells of the nodules. Area of interactions between T- and B-cells.

3. Splenic circulation. Important for understanding splenic function.

a) **Splenic artery** enters the hilus and branches in trabeculae as **trabecular arteries**. These give rise to **central arteries** (actually arteriole) which exit trabeculae and are surrounded by periarteriolar lymphatic sheath of T-lymphocytes (PALS) and come to be eccentrically located in a splenic nodule.

b) Central arteries give rise to smaller branches which ultimately open into or terminate in pulp cords = **open circulation**. Blood can be "cleansed" by macrophages of pulp cords and make its way back into the circulation through the walls of venous sinusoids.

c) **Venous sinuses** have loosely organized, longitudinally oriented endothelial cells (non-phagocytic) which are surrounded by a very leaky basal lamina; thus cells can readily move into and out of the sinus to pulp cords. Venous sinuses drain to venules which drain to pulp veins which drain to trabecular veins which empty into the splenic vein.

4. Thus, one function of the spleen is carried out in pulp cords, i.e., filtration of antigens in blood and removal of worn-out RBCs and platelets. Hemoglobin of phagocytosed RBCs is broken down and various components reused as appropriate.

a) Other functions. Some lymphopoiesis occurs in the spleen and antibodies are secreted by plasma cells and B-cells.

C. **Thymus.**

1. General function. Site of lymphocyte proliferation and transformation of lymphocytes into immunologically compentent T-lymphocytes.

2. Structure. 2 lobes, no lymph nodules, no lymph sinuses, no afferent lymph vessels (does not filter lymph). Has a **cortex** and **medulla**.

a) Stroma. Consists of **thymic reticular cells** joined by desmosomes to form a meshwork of spaces.

b) Parenchyma. Consists of lymphocytes and macrophages occupying the meshes of the thymic reticular cells. Thymic reticular cells differ from reticular cells of other lymphatic tissue in that they are endodermal in origin, epithelial in character and produce a hormone-like lymphokine = **thymosin**.

c) **Cortex**. Rich in lymphocytes; are immature and not yet immunologically competent. **Blood-thymic barrier** is in the cortex and consists of capillaries surrounded by thymic reticular cells. Barrier prevents antigens from prematurely contacting developing lymphocytes. Components of the barrier include the capillary endothelium, its basal lamina, perivascular connective tissue space, basal lamina of thymic reticular cells and thymic reticular cell cytoplasm.

    d)   **Medulla**. Fewer lymphocytes and more thymic reticular cells than in cortex; lymphocytes are immunologically competent and ready to exit the thymus and interact with a specific antigen.

    e)   **Thymic corpuscles** or **Hassall's corpuscles**. Present in the medulla. Composed of concentric layered whorls of degenerate thymic reticular cells.

  3. Function of thymus. Lymphopoiesis and transformation of lymphocytes into immunologically competent T-lymphocytes that can respond to specific antigens. Transformation requires interaction with thymic reticular cells, and involves the production of a hormone-like substance, thymosin, by reticular cells.

    a)   Transformed T-lymphocytes exit thymus from the medulla by diapedesis into post-capillary venules, then seed or colonize thymus-dependent areas of lymphatic tissues and organs.

    b)   Activity. Most active during the neonatal period through puberty with a postpubertal decline in activity.

    c)   Hormonal influences. Sex steroids and adrenal steroids (and stress) depress thymic activity. This could relate to postpubertal decline in thymic activity. Thyroxine favors growth and development of the thymus.

MUSCLE TISSUE

I.   An organization of contractile cells which produces movements, e.g., to close an opening (sphincter), propel material (blood, GI or genitourinary systems), move a joint and provide locomotion.

  A.   Features. Elongated cells = fibers; present as sheets, bundles or fascicles. Cells (fibers) are excitable and influenced by nerves and hormones.

  B.   Types: striated (skeletal and cardiac) and smooth.

II. **Skeletal muscle** (striated and under voluntary control).

  A.   Light microscopic level of organization.
    1. Fiber (myofiber, cell).
      a)   Contains multiple nuclei which are peripherally located. **Satellite cells** are closely adjacent to myofibers; these are important in repair and regeneration of skeletal muscle. Fibers are organized into fascicles (groups of fibers) and into whole named muscles by connective tissue.

        (1) **Epimysium**. Dense connective tissue that invests muscle as fascia. This connective tissue investment inserts into bone via tendon to anchor the muscle.

        (2) **Perimysium**. Invests fascicles as functional units.

        (3) **Endomysium**. Basically is a basement membrane plus fine collagenous connective tissue to promote connections between adjacent myofibers. Satellite cells lay within the endomysium.

        (4) Nerves and blood vessels use the connective tissue investments as avenues to ramify and supply the muscle.

b) Cell membrane = **sarcolemma**. Is excitable. Has invaginations = **transverse tubules** (T-tubules) for transmission of depolarization wave into the interior of the fiber. Sarcolemma is surrounded by an external lamina = basal lamina.

c) **Myofibrils**. Evident as the 'meat' of each fiber. Composed of **myofilaments**, actin and myosin, which are organized to form repeating units called **sarcomeres**. Registration of sarcomeres produces the striations of skeletal muscle.

B. Electron microscopic level of organization

1. **Sarcomere**. Composed of the myofilaments: actin (thin) and myosin (thick) within a myofibril. Within a sarcomere there is a **banding pattern** due to the organization of the myofilaments: A-band, I-band, H-zone and Z-line (Z disc).

   a) Two Z-lines delineate a sarcomere.

   b) Thick filaments are limited to the A-band.

   c) Thin filaments extend from Z-line forming the I-band and interdigitate with thick filaments in the A-band.

   d) H-zone: clear area of A-band where thin filaments are lacking.

2. Organelles. Skeletal muscle is rich in mitochondria (energy production as ATP for contraction) and sarcoplasmic reticulum (sER, involved in Ca^{++} release/sequestration important for muscle contraction and relaxation). Glycogen is moderately abundant (glucose stores). Few rER or ribosomes are present as protein synthesis is limited; cell is specialized for contraction.

3. **Triads**. Membrane specializations in the interior of the striated fiber. Important for depolarization and Ca^{++} release for contraction.

   a) Components. Transverse tubule (**T-tubule** = invagination of the sarcolemma) and two expanded **terminal cisternae** of sER (sER = network of membrane over and around myofibrils).

   b) Location. Junction of A-I bands of myoribrils in skeletal muscle.

   c) Function. The neurotransmitter acetylcholine binds to sarcolemmal receptors which leads to depolarization. Depolarization wave travels to interior of myofiber via T-tubule and causes release of Ca^{++} from terminal cisternae of sER which leads to initiation of sliding of myofilaments of myofibrils = contraction, with resultant changes in bands of sarcomere.

4. Other components of cytoplasm (sarcoplasm).

   a) **Myoglobin**. An oxygen binding protein. Varies in amount between the different **types of myofibers**, red vs. white vs. intermediate.

      (1) **Red**. High content of myoglobin (gives dark color to muscle), muscle capable of sustained activity over time (slow twitch). Also many mitochondria.

      (2) **White**. Little myoglobin (light colored fibers). Muscle produces short bursts of activity (fast twitch). Few mitochondria but rich in glycogen.

      (3) **Intermediate**. Characteristics between red and white. Most human muscles are a mixture of red, white and intermediate.

C. Specialized structures of skeletal muscle.

1. **Myotendinous junctions**. Connection of muscle to tendon. Collagen and reticular fibers of the tendon insert into the external (basal) lamina of the myofiber to strongly bind muscle and tendon together.

2. **Golgi tendon organs**. Stretch receptors of tendon. Activated by tension on tendon, afferent fiber carries information to spinal cord to inhibit neurons that innervate the muscle. Prevents tearing and damage to the muscle.

3. **Myoneural (neuromuscular) junction (motor end plate)**. Axon of **alpha** motor neuron forms a synaptic junction to innervate a myofiber. Releases acetylcholine for activation of receptors and contraction (i.e., neuromuscular transmission).
   a) Motor unit. One axon plus all of the muscle fibers it innervates. Once the axon fires, all innervated myofibers contract as a unit. Ratio of axon:myofibers determines the level of control of a muscle; intrinsic muscles of the hand have lower ratio than deep back muscles.
4. **Neuromuscular spindle**. Involved in muscle tone; number increases in muscles that are finely controlled (hands). Contain encapsulated **intrafusal** muscle fibers (**extrafusal** fibers are the regular fibers of the muscle). Innervated by **gamma** motor neurons. Also receive afferent (sensory) nerves. Gamma motor neurons keep the intrafusal fibers partially contracted (= tone) so any movement of extrafusal fibers changes this tone and causes the afferent fibers to fire and send information to the spinal cord (CNS) so adjustments can be made.

D. Regeneration and repair. Myofibers do not divide; but satellite cells are stimulated by damage to divide and differentiate into skeletal muscle fibers. Hypertrophy of a "muscle" due to strenuous excercise results from an increase in myofibrils in myofibers and increased numbers of myofibers differentiating from satellite cells.

III. **Cardiac muscle** (striated and not under voluntary control).

A. A.Light microscopic level of organization.
   1. Fiber (myofiber, cell). Contains a single nucleus that is centrally located. Fibers are branched and their branching arms are joined by specialized junctions = **intercalated disks.** These junctions have mechanical and electrical components.
   2. Muscle is profusely vascularized.
   3. Fibers differ between atria and ventricles. Atrial fibers: smaller and possess secretory granules containing a hormone = **atrial natriuretic peptide** (atriopeptin).

B. Electron microscopic level of organization.
   1. Similar to that of skeletal muscle; however, there are exceptions.
      a) Sarcoplasmic reticulum (sER) is not as well developed and terminal cisternae are not as expanded.
      b) Interactions of one terminal cisternae of sER and a T-tubule produces a **diad** (rather than a triad as in skeletal muscle).
      c) Diad is placed at Z-line (triad of skeletal muscle is at A-I junction).
   2. Inclusions: lysomsomal material accumulates with age as lipofuschin pigment (cardiac muscle cells do not undergo mitosis to redistribute accumulating pigment). This eventually may interfere with normal functions.
   3. **Intercalated disk**. 2 parts and 3 components.
      a) Parts. Transverse part oriented transversely to long axis of the fiber and longitudinal part oriented parallel to long axis of the fiber.
      b) Components. **Gap junction** (nexus) at the longitudinal component; electrical junction for communication and coordination of contractions. **Fascia adherens** at the transverse component; sites of insertion of actin filaments at the terminal sacromere of the fiber and transmits contractile forces to connected fibers. **Macula adherens** (desmosome) at the transverse component; mechanically holds fibers together to prevent separation during contraction.

C. Innervation. Supplied by sympathetic, parasympathetic and sensory nerve fibers. No direct innervation or specialized junction between nerve and muscle cell; axons travel close to myofibers, transmitter is released and diffuses over a relatively long distance to reach sarcolemma. Excitation is transmitted to adjacent myofibers via gap junctions of the intercalated disks.

D. Regeneration and repair. Cardiac muscle does not regenerate. Damaged muscle is replaced by scar tissue (fibroblasts and extracellular fibers).

IV. **Smooth muscle** (not striated and not under voluntary control).

A. Light microscopic level of organization.
   1. Fiber (cell). Contains a single centrally located nucleus. Fibers are long and tapered. They form bundles or sheets. Organelles are perinuclear. There is no visible evidence of striations.

B. Electron microscopic level of organization.
   1. Myofilaments of actin and myosin are present, but not organized as myofibrils. Myosin filaments are considered labile and not normally seen except immediately prior to contraction.
   2. **Dense bodies** (alpha-actinin) are seen throughout the cytoplasm and along the sarcaolemma (modified Z-lines). Dense bodies serve as sites for insertion of actin filaments. During contraction, the interaction of myosin, actin and dense bodies produce a ruffling (shortening) of the fiber
   3. T-tubule and sarcoplasmic reticulum (sER) system does not exist. **Cavolae** (invaginations of the sarcolemma) may be rudimentary T-tubules. Subsarcolemmal smooth membranous vesicles may be rudimentary sER.
   4. Smooth muscle contraction works through a Ca^{++}/calmodulin system, unlike skeletal and cardiac muscle, thus sER may not need to be well developed.

C. Innervation. Supplied by sympathetic, parasympathetic and sensory nerve fibers. No direct innervation or specialized junction between nerve and muscle cell. Axons travel close to myofibers, transmitter is released and diffuses over a relatively long distance to reach the sarcolemma. Excitation is transmitted to adjacent myofibers via gap junctions.
   1. Two types of smooth muscle with regard to innervation. **Multiunit** smooth muscles are richly innervated which gives fine control (e.g., ciliary muscles of the eye). **Visceral or unitary** smooth muscles are poorly innervated, rely on connections via gap junctions to spread excitation, muscle is poorly controlled and works in sheets (e.g., myometrium of uterus, muscularis of ureter).

D. Hormonal interactions. Many smooth muscles respond to hormones, e.g., uterine smooth muscle responds to oxytocin; GI smooth muscle responds to cholecystokinin, motilin and serotonin; and vascular smooth muscle responds to epinephrine.

E. Regeneration and repair. Smooth muscle cells can undergo modest cell division and repair when damaged.

NERVE TISSUE see section on Neuroanatomy.

CARDIOVASCULAR SYSTEM

I. Circulatory system. Comprises heart, blood vessels (arteries, veins and capillaries) and lymphatic vessels. General function: carry O_2, nutrients and wastes from one location in the tissues and body to another; also to distribute cells of the defense system and hormones and drain fluids from tissues.

II. Cardiovascular system (CV). Comprises heart and blood vessels.

   A. Two circulations are the systemic and pulmonary. Both use heart, arteries, arterioles, capillaries, venules and veins. Vessels tend to travel as companions, e.g., arteries with veins, arterioles with venules and precapillary arterioles with postcapillary venules.

   B. **General principle**. The arterial system is a higher pressure system than the venous sytem, therefore the arterial vessel wall is thicker and the lumen is narrower relative to the companion vessel of the venous system. Venous system is a lower pressure system with vessels of thinner wall, but larger diameter lumen.

   C. Element common to entire CV system is: the lining of all components is simple squamous epithelial cells which constitute an **endothelium**. A basement membrane underlies this endothelium. All elements entering or exiting the CV system must cross the endothelium and basement membrane.

   D. General histology of the CV system. Three tissue layers (tunics) that are modified in size and constituents depending on the particular vessel's function.
      1. **Tunica intima** (TI) (= endocardium of heart). Components are endothelium (simple squamous endothelial cells that are linked by a mixture of gap and occluding junctions; junctions are in arteries only, not in capillaries and rare in veins), basement membrane, subendothelial loose connective tissue, and internal elastic membrane.
      2. **Tunica media** (TM) (= myocardium of heart). Components include smooth muscle (circular arrangement) mixed with elastic fibers or elastic laminae, and external elastic membrane. Control of luminal diameter rests here.
      3. **Tunica adventitia** (TA) (= epicardium of heart). Components are loose connective tissue containing some smooth muscle cells and adipocytes. Some nerves and blood vessels (nervi and vasa vasorum, respectively) in the adventitia penetrate to the media of larger vessels. TA serves to anchor vessels to surrounding tissues.

III. Specific vessels.

   A. **Elastic arteries**. Examples include the aorta, carotid, pulmonary arteries. Many elastic laminae mixed with smooth muscle in the TM. Elastic tissue stretches during systole, stores potential energy for recoil during diastole (promotes continuous flow of blood in the capillary, rather than intermittent with each heartbeat).

   B. **Muscular (distributing) arteries**. Include named arteries, e.g., radial, femoral, basilar and uterine. Thick wall relative to luminal diameter. Prominent internal and external elastic laminae; but there is a progressive shift in proportion of smooth muscle relative to elastic fibers in TM (toward increase in smooth muscle; 6-40 layers of myofibers). Muscle responds to autonomic nerves and hormones and therefore participates in regulating blood flow to an area. Prominent TA.

C. **Arterioles** (smallest arteries). Are sphincters and blood pressure regulators. Thick wall relative to diameter of lumen. Internal elastic lamina present in larger arterioles, but absent in smaller arterioles. TM has 1-5 layers of smooth muscle. Smallest arterioles (**metarterioles**) may have non-muscle cells (pericytes) surrounding lumen; metarterioles are continuous with capillaries. Metarterioles can reduce blood flow to a vascular bed. A single smooth muscle cell may surround a capillary at its origin from the metarteriole, this is the **precapillary sphincter** which can shut-down blood flow to the capillary bed.

D. **Capillaries.** Consist of a single layer of endothelial cells plus the basement membrane. Two main types of capillaries.
   1. **Continuous.** Most common and are located in skeletal muscle, brain, dermis of skin, lungs and salivary glands. Endothelial cells are joined by 'spots' of occluding junctions. Fluid (water) can cross the space between these junctions, but protein macromolecules cannot. Endo- and pinocytotic vesicles are used for transport of large molecules. Oxygen and glucose can diffuse freely across the capillary wall.
   2. **Fenestrated.** Located in endocrine glands , renal glomerulus and mucosa of the intestine (sites where rapid exchange of materials is desirable). Endothelial cytoplasm is attenuated and marked by numerous 'holes' or fenestrae. Fenestrae are closed by a **diaphragm** (thinner than a cell membrane). They allow rapid fluid exchange. **NOTE:** renal glomerular fenestrated capillaries lack diaphragms, but a thick basement membrane acts as a selective filter/barrier.
   3. Third type of capillary is often referred to as a **discontinuous capillary or sinusoid.** Present in liver, red pulp of spleen and in loose connective tissue as lymph capillary. Characteristics include gaps between endothelial cells, discontinuous or thin basement membrane and large, irregular shaped lumen. Even cells can pass across this capillary.

E. **Postcapillary venules.** Loosely organized endothelium, large gaps between cells, thin to incomplete basal lamina, receptors for many vasoactive pharmacologic agents on endothelial cells (e.g., histamine that makes these vessels leaky during inflammatory reactions). True TM lacking; lumen is surrounded by a loose arrangement of filament-containing pericytes. TA is very thin. All of these features contribute to these vessels being "leaky" and involved in **blood-interstitial fluid exchange** and extravasation of water and solutes (edema) and **diapedesis** of blood cells. **NOTE:** the postcapillary venule is the site where cells can exit the vascular system and enter the connective tissue.

F. **Venules.** Have a continuous endothelium, fluid does not exit these vessels. TM may have a layer or two of smooth muscle. Lumen is large relative to the thickness of the wall. TA merges with surrounding connective tissue. Parallel arterioles.

G. **Veins.** Usual TI. TM may have several layers of muscle mixed with some elastic fibers. TA prominent. Lumen large relative to wall thickness. Valves in these veins = fold of TI; prevent reflux of blood. Valves are especially present in veins of the lower extremities. **NOTE:** blood may pool in sinus of valve = varicose veins.

H. **Large veins.** TI is usual. TM may have longitudinal muscle (veins of limbs). TA may have longitudinal muscle (e.g., abdominal veins such as the inferior vena cava). TA usually prominent. Lumen quite large, but wall is thin.

IV. Heart. Basically a muscular pump, but the atrium is also an endocrine organ. Atrial muscle secretes atrial natriuretic peptide (atriopeptin; involved in sodium balance and blood pressure regulation). Three layers like vessels, but are modified for the pump.

A. **Endocardium** (= TI). Endothelial layer, subendothelium containing connective tissue with collagen and elastic fibers and scattered smooth muscle. Purkinje fibers (part of the impulse conduction system) are present in the subendocardium of the ventricles.

B. **Myocardium** (= TM). Consists of striated cardiac muscle. Myofiber size and content varies with regions. Atrial fibers smaller than ventricular fibers; atrial wall is thinner than ventricular wall; left ventricular muscle is the thickest.

C. **Epicardium** (= TA; also forms the **visceral pericardium**). Lined by mesothelium (a wet, serous membrane). Connective tissue underlying the epithelium is unremarkable except it contains adipose tissue. Large (coronary) blood vessels and nerves travel in this area.

D. **Cardiac skeleton**. Dense fibrous connective tissue. Forms the annuli fibrosi (surrounds the atrioventricular openings) and the fibrous trigone and septum membranaceum (of the interventricular septum). Serves as sites of attachment of cardiac muscle.

E. **Valves**. Consist of a core of dense connective tissue continuous with the annuli fibrosi. Core is covered on both sides by endocardium. Chorda tendinae attach to the valves and are also covered by endocardium. The valves are the mitral (bicuspid) (left atrioventricular valves), tricuspid (right atrioventricular valves), pulmonary and aortic valves. **NOTE**: no blood vessels in valves; do not want inflammatory reactions here which could alter morphology and thus function of the valves.

F. **Impulse conducting system**. Coordinates heartbeat, rate and rhythm. Cardiac muscle here is modified from the 'working' myocardium. **Sino-atrial (SA) node** (initiates heartbeat/pacemaker) is located at junction of superior vena cava and right atrium; muscle fibers are smaller and with fewer myofibrils, and muscle is mixed with dense connective tissue. Fibers of SA node connect with the **atrio-ventricular (AV) node** located in the interatrial septum (lower part). Structure of AV node is similar to the SA node. Fibers from the AV node are continuous with the **atrio-ventricular (AV) bundle** (right and left bundle branches) that course in the interventricular septum. Fibers of the bundle branches are continuous with **Purkinje fibers** which continue into the papillary muscles. Purkinje fibers are larger than other myocardial cells, have peripherally located myofibrils, and a central store of glycogen and a central nucleus. Fibers of the conduction system communicate via intercalated discs. Conduction system is innervated with sympathetic and parasympathetic nerves; these nerves modify heart rate, rather than initiate heartbeat. **SA node is the pacemaker.**

G. Chemoreceptors. **Carotid body**. Located at the bifurcation of the common carotid artery. Is richly vascularized and contains special catecholamine-containing cells innervated by afferent fibers. Senses O_2 and CO_2 tension, and pH. Afferent information carried to the CNS alters cardiovascular parameters.

H. Baroreceptors. **Carotid sinus**. In the bifurcation of the carotid arteries. Consists of afferent nerves that respond to stretch (pressure). Afferent information carried to the CNS alters cardiovascular parameters.

V. Age changes in the CV system. Decrease of elasticity and elastic tissue in large arteries. Loss of structure of the arterial wall leads to ballooning of the wall = **aneurysm**. Infiltration of TI with connective tissue elements and lipid = **atherosclerosis**. Increase in lipofuscin in cardiac muscle which impacts muscle function. Calcification of cardiac skeleton.

INTEGUMENTARY SYSTEM

I. Components: skin and its appendages/derivatives (nails, hair, sweat and sebaceous glands). Skin functions as a barrier to infections and dehydration, is important for sensory reception and temperature regulation, filters UV light (melanocyte system), and participates in excretion (sweat glands). Skin is invaluable in wound healing; epidermis and epidermal derivatives are capable of remarkable re-epithilization after simple or marked damage.

II. Skin.

   A. Types of skin: thick and thin.
   1. Thick skin. Examples are on the palms of hand and sole of foot. Thickness of 0.8 to 1.5mm. Is subjected to more mechanical shearing forces and abrasions than thin skin.
   2. Thin skin. An example is the skin over the general body including the abdomen and back. Is 0.1 to 0.5mm thick.

   B. Components: epidermis and dermis.
   1. **Epidermis** is of ectodermal origin. Epithelial type = stratified squamous keratinized. It is avascular and contains free nerve endings for pain perception. It rests on a basement membrane.
   2. **Dermis** is of mesodermal origin. Is dense irregular connective tissue that is vascular and contains encapsulated nerve endings (Pacinian and Meissner's corpuscles).
       a) Dermis is comprised of 2 layers: papillary (nearest epidermis and most vascular) and reticular (deeper and more dense connective tissue).
       b) Dermal projections from papillary layer (dermal papillae) interdigitate with epidermal projections (rete ridges or epidermal pegs). These interdigitations help adhere epidermis to dermis.

III. Epidermis. Comprised chiefly of cells called keratinocytes. Cells form layers which are (from deepest near dermis to most superficial):

   A. **Stratum basale** (stratum germinativum). Single layer of cuboidal cells abutting the basement membrane. Is attached through the basement membrane to the dermis by hemidesmosomes. This is the proliferative layer, cells become keratinocytes and replenish (regenerate) upper layers.

   B. **Stratum spinosum**. Variable number of cell layers (from one to six depending on the thickness of the skin and body part). Cells accumulate tonofilaments that insert into **desmosomes** - junctions that mechanically couple the cells.

   C. **Stratum granulosum.** Three to five layers of cells. Accumulate keratohyaline granules (produces basophilic staining). Keratinocytes also contain packed tonofilaments and **lamellar granules** (membrane-coating granules) that discharge their contents of GAGs, phospholipids and glycolipids into the extracellular spaces. This material produces a 'sealing' material or barrier to penetration of materials through the epidermis.

   D. **Stratum lucidum**. Is a translucent layer most prominent in thick skin. Cells are anucleate and lack organelles. Filaments packed in cytoplasm. Desmosomes are prominent.

E. **Stratum corneum**. Five to ten layers of packed, cornified cells (plates). They are filled with filaments embedded in an amorphous material contributed by the keratohyaline granules. Cells are attached by desmosomes except at the uppermost layers where the cells tend to shed. These cells produce a barrier to limit loss of water from the dermis and to prevent entry of noxious material from the exterior.

F. **Stratum disjunctum**. Term often applied to the layers of keratinocytes that are constantly being shed. Look for them in your bed.

G. Process of differentiation of keratinocytes from stratum basale to stratum corneum is 15-30 days. Parallels the time course for wound healing.

IV. Cells of the epidermis.

A. **Keratinocytes**. Epithelial cells that undergo differentiation or keratinization.

B. **Melanocytes**. Neural crest derived cells found mainly in the stratum basale and in hair follicles. Produce the pigment **melanin**. Melanin granules are dispersed in long processes of the melanocyte and then transferred to keratinocytes of the strata basale and spinosum. Produces coloration of the skin and hair (also found in the iris of the eye). **NOTE**: melanin is deposited supranuclearly in keratinocytes and serves to protects the DNA of the proliferative keratinocytes from UV radiation.

C. **Langerhans cells**. Cells of the immune sytem (dendritic cells, antigen presenting cells) located primarily in the stratum spinosum. Present antigens to T-lymphocytes of epidermis. Contain unusual granules called Birbeck granules (function unknown).

D. **Merkel cells**. Located in stratum basale. Many dense-core granules in the cytoplasm. Innervated by nerve fibers - involved in sensory reception; tactile receptors.

V. Derivatives of the epidermis.

A. **Nails** = nail plate. Nail plate is essentially the stratum corneum and covers the **nail bed** (strata basale and spinosum). Nail developes from the stratum basale of the **nail matrix** (most proximal part of the bed). **NOTE**: if the nail matrix is damaged, the nail does not regenerate. Nails are transparent and can be a window on the degree of oxygenation of underlying vascularized tissue.

B. **Hair**. Invagination of the surface epidermis gives rise to the hair follicle. Expanded end of the follicle is the **hair bulb** containing the **matrix** (equivalent to the stratum basale). A vascularized connective tissue tuft invaginates the bulb as the **dermal papilla**. The dermal papilla induces division of the epithelial cells of the matrix and thus growth of the hair shaft. Damage the dermal papilla and the hair does not grow. Hair color is due to melanocytes in the matrix. The process of keratinization in the hair is similar to that of the covering epidermis.

C. **Sebaceous glands**. Located in the dermis, but attached to hair follicles. Produce a lipid secretion (sebum) that is released by holocrine secretion (whole cell is released and is broken down). Release occurs along hair shafts. Sebum may participate in the hydrophobic barrier of the skin.

D. **Sweat glands**. Located deep in the dermis. Simple coiled tubular glands. Ducts traverse the dermis and epidermis to open at the suface.

1. **Eccrine.** Produce sweat which is hypotonic containing water, sodium and chloride. On evaporation, the body cools. These secretory cells are innervated by cholinergic autonomic nerves. Myoepithelial cells surround the secretory cells. Located on general body surface.

2. **Apocrine.** Located near anus, axilla and areola of nipple. Secretory units are highly expanded and surrounded by myoepithelial cells. Secretion is viscous and becomes odoriferous on bacterial action. Secretion influenced by sex steroids and sympathetic adrenergic nerves.

E. **Mammary glands** (see female reproductive system).

VI. Nerve endings associated with skin.

A. Sensory (afferent) nerves. Dendritic endings of dorsal root ganglion cells.
1. Free nerve endings. Occur in the epidermis. Exist as naked dendritic endings of dorsal root ganglion cells. Nociceptors (respond to pain).
2. Meissner's corpuscles. Located in dermal papillae. Are encapsulated endings that sense light touch.
3. Pacinian corpuscles. Are deep in the dermis. Encapsulated endings that sense deep pressure and vibration.
4. Merkel discs. Present in the epidermis. Nerve endings on Merkel cells are receptive to pressure.
5. Peritrichial nerves. Occur around hair follicles in the dermis. Sense light touch and movement of the hair.

B. Motor (efferent) nerves. Sympathetic postganglionic axons. Innervate arrector pili smooth muscle associated with hair follicles (noradrenergic axons). Innervate arteries (noradrenergic axons) and innervate sweat glands (cholinergic axons).

RESPIRATORY SYSTEM

I. System designed to take in air and modify it to accomplish respiratory gas exchange, olfaction and sound production. Air is modified by cleansing, moistening, warming or cooling.

II. Two components of the system: air conduction portion (no gas exchange) and respiratory portion (respiratory gas exchange).

A. **Conducting portion.** Includes the nasal cavity, nasopharynx and oropharynx, larynx, trachea, bronchi and bronchioles.

B. **Respiratory portion.** Includes the respiratory bronchioles, alveolar ducts, alveolar sacs and alveoli.
1. The **blood-air barrier** (respiratory membrane) is in the respiratory portion. (See IV.F. below for description).

III. Conducting passages. Lining epithelium is "**respiratory epithelium**" = pseudostratified ciliated columnar with goblet cells (except bronchioles which are lined by ciliated columnar epithelium). Mucus from goblet cells and cilia help trap particulate matter and move it to the exterior.

A. Nasal cavity. Contains 3 conchae which increase surface area and create turbulence to aid warming, moistening and cleansing air (particles are trapped in moist nasal mucosa). Respiratory epithelium lines the cavity (cilia beat to move trapped particles into the oral cavity). Lamina propria (loose connective tissue underlying the epithelium) contains mucous glands, serous glands and a venous plexus (warms air). **NOTE**: the venous plexus can become engorged (swollen) and "leaky" in allergic reactions; blocks nasal passages.

B. **Olfactory epithelium**. Located in the superior aspect of the nasal cavity and consists of pseudostratified ciliated columnar epithelium without goblet cells. Three cell types: olfactory receptor cells, supporting cells and basal cells (stem cells).
   1. Receptor cell. **Bipolar neuron** whose dendrite extends to the surface and ends in a ciliated (non-motile) olfactory vesicle which is responsible for signal transduction of odors. An axon extends from the bipolar cell body into the olfactory bulb of the CNS.
   2. Lamina propria contains **Bowman's glands**. Serous glands whose ducts enter the epithelium and carry a product to the surface which acts as a solvent for odorous substances.

C. **Nasopharynx**. Connects nasal passages to larynx. Epithelial lining is mainly respiratory epithelium, but some stratified squamous nonkeratinized near oropharynx and where there is some "wear and tear". Pharyngeal tonsils (adenoids) are in posterior part of nasopharynx.

D. **Trachea**. Conducts air to bronchi. Hyaline cartilage, c-shaped 'rings' keep it patent. Wall comprised of 4 layers:
   1. Mucosa. Lined by 'respiratory epithelium' resting on a thick, prominent basement membrane and a lamina propria (loose connective tissue layer subjacent to most epithelia lining a lumen). Epithelial cell types: **goblet cell** (unicellular, mucous gland), **ciliated cell**, **basal cells** (stem cells), **brush cell** (have apical microvilli; some are stem cells, other are innervated and may be sensory receptors), and **small granule cells** (contain catecholamines and may coordinate functions of other epithelial cells). Free nerve endings in epithelium respond to irritants.
   2. Submucosa. Connective tissue more dense than lamina propria. Extends from lamina propria to perichondrium of cartilage. Contains seromucous glands.
   3. Muscularis. Contains cartilage ring anteriorly and smooth muscle (trachealis) posteriorly.
   4. Adventitia. Connective tissue extending from perichondrium to adjoining structures. Contains blood and lymph vessels and nerves.

E. **Bronchi**. Similar wall structure to that of trachea. Cartilage 'rings' are replaced by **cartilage plates**. Moving distally in the bronchial tree, cartilage decreases (seen as irregular plates) and smooth muscle increases, glands decrease in number and epithelium becomes lower.

F. **Bronchioles**. Lack cartilage and seromucous glands. Epithelium is simple columnar to cuboidal with scattered goblet cells, but ciliated cells present. Clara cells are present in epithelium—may produce a surfactant-like material and may play a role in moisturizing the bronchiolar lumen. **NOTE**: throughout the respiratory system, elastic fibers are numerous and important; they allow expansion and aid in expiration of air and they tether the bronchiolar wall to lung parenchyma to keep it patent. In emphysema, elastic tissue is reduced (by elastase from macrophage lysosomes), thus, bronchioles tend to collapse on expiration leading to inefficient respiration.

IV. Respiratory passages. Sites of blood gas exchange.

A. **Respiratory bronchiole**. Alveoli pouch off the bronchiolar wall. Smooth muscle is present in the wall between pouching alveoli. Ciliated cells in epithelium except in alveolar wall.

B. **Alveolar duct**. Wall composed of alveoli. Minimal smooth muscle in wall. Open into blind sacs = alveolar sacs.

C. **Alveolar sac**. Cul-de-sac composed of alveoli. Lacks smooth muscle. Lined by simple squamous epithelium. Only connective tissue is elastic fibers and a few reticular fibers.

D. **Alveolus**. Pouch-like structure with a wall composed of 2 cell types.
   1. **Aleolar type I** (type I pneumocyte). Simple squamous epithelial cell which covers about 97% of the surface area of the alveoli.
   2. **Alveolar type II** (type II pneumocyte; great alveolar cell). Bulges into air space and are secretory cells. They contain multivesicular bodies or cytosomes rich in phospholipids and GAGs and produce **surfactant** which lines alveolar spaces. Connected to type I cells via junctional complexes.
   3. **Alveolar macrophages** (dust cells). **Not part of the epithelium**, but wander over epithelium to cleanse alveolar spaces by phagocytosing inhaled particles. They are moved to exterior by adherance to mucus and transport via cilia.

E. Interalveolar wall (septum). Separates the alveolar air space of 2 adjacent alveoli. Components of the wall are Type I cells (bordering each air space) and connective tissue intervening between the type I cells. Connective tissue contains capillaries, fibroblasts, elastic fibers and monocytes. **NOTE**: monocytes give rise to the alveolar macrophages. Interalveolar pores are 10-15μm diameter and connect air spaces of adjacent alveoli. Allows equilibration of air pressure and movement of macrophages to adjacent alveoli.

F. **Blood-air barrier** (respiratory membrane - of all alveoli and interalveolar septa). Consists of type I alveolar cell + its basal lamina) <u>and</u> capillary wall + its basal lamina.

G. Primary pulmonary lobule. A single respiratory bronchiole + many alveolar ducts + many alveolar sacs + numerous alveoli.

V. Blood supply. Oxygen-poor blood from the right heart ventricle is distributed by pulmonary arteries to pulmonary capillaries (oxygen-poor blood from the systemic bronchial arteries is mixed in the capillary system). Oxygen-rich blood is carried from pulmonary capillaries to pulmonary veins to left heart atrium to enter systemic circulation.

VI. Innervation. Parasympathetic via the vagus nerve; vagal stimulation causes bronchial constriction. Sympathetic stimulation causes bronchial dilation. **NOTE**: sympathomimetic drugs are used to dilate the bronchial system during asthma attacks. Sensory nerves of respiratory system may be nociceptors and signal reflexes.

GASTROINTESTINAL SYSTEM

I. Components:

   A. **Upper GI system**. Includes the oral cavity (lips, tongue, teeth, oral mucosa and salivary glands), and pharynx. Functions to begin mechanical breakdown of ingested food, initiate chemical digestion (carbohydrates) and transport materials to the stomach.

   B. **Lower GI system**. Includes the esophagus, stomach, small intestines (duodenum, jejunum and ileum), large intestine (colon & appendix), rectum and anal canal. Function: to transport ingested material to provide a site for enzyme digestion (stomach and small intestine) which is aided by liver, pancreas and gall bladder. This results in the formation of chyme—semi-fluid material in which lipids, proteins and carbohydrates are broken down into constituent

189

parts for absorption in small intestine. Absorption of water and electrolytes occurs in the colon which results in compaction of fecal material.

C.   **Glandular derivatives** of the intestine are the pancreas, liver and gall bladder.

II.   Upper GI system.

A.   **Oral cavity**. Boundaries = lips, cheeks, hard and soft palates and floor of the mouth. Contains special structures of teeth and tongue.
1.   **Oral mucosa**. Wet mucous membrane consisting of epithelium (stratified squamous epithelium) and lamina propria (loose connective tissue with elastic fibers). (**NOTE**: inner lip, soft palate, and buccal mucosa are covered by stratified squamous non-kertinized epithelium, but hard palate, gingiva and superior surface of tongue are covered by stratified squamous keratinized epithelium). Mucosa is wetted by secretions of salivary glands.
2.   **Lips**. Essentially a core of skeletal muscle (orbicularis oris) surrounded by fibroelastic connective tissue and covered by skin on the outer surface and oral mucosa on the inner surface. Abrupt change of epithelium (kertinized to non-keratinized) at the **vermillion border**. Minor salivary (mucous) glands in the oral mucosa.
3.   **Cheeks**. Essentially the same structure as the lips. Muscle and fibroelastic connective tissue of lips and cheek function in facial expression and phonation, as well as in mastication.
4.   Palate.
a)   Hard. Forms roof of mouth and has a core of bone. Food is thrust against this surface during mastication, therefore, oral mucosa is firmly attached to the periosteum of bone and covered by keratinized epithelium. Mucous glands reside in the posterior 2/3 of the hard palate.
b)   Soft. Continues from posterior part of hard palate and joins with uvula. Core of skeletal muscle, fibroelastic connective tissue and seromucous glands. Epithelium - oral side is stratified squamous non-keratinized; nasal side is respiratory epithelium.
5.   **Salivary glands**. **Minor** glands are scattered throughout the oral cavity; mainly mucous type, though lingual are seromucous. **Major** glands are the parotid, submandibular and sublingual.
a)   Parotid (serous acini). Serous secretion is watery and proteinaceous and contains digestive enzymes (e.g., amylase, maltase and ptyalin). IgA immunoglobulin produced by parotid gland protects against orally ingested pathogens.
Cells. Serous secretory acinar cells (prototype of a protein synthetic cell) abutted by myoepithelial cells. Striated duct cells alter the glandular secretion.
b)   Submandibular (submaxillary)(mixed seromucous acini). Serous elements present as demilunes. Serous cells produce an enzyme, lysozyme, that hydrolyzes the walls of bacteria.
c)   Sublingual (mainly mucous, but some serous acini). Serous elements present as demilunes. Serous cells produce lysozyme.
d)   Saliva. Major constituents are water, glycoproteins, enzymes, immunoglobulins, proteoglycans and salts. Water and glycoproteins serve a lubrication function.
e)   Innervation of salivary glands. Parasympathetic nerve stimulation results in abundant watery fluid containing little organic products. Sympathetic nerve

stimulation results in a small amount of viscous material rich in organic products ('dry mouth' of fight or flight response).

6. **Tongue**. Core of skeletal muscle organized in 3 planes (for fine control and movement). Muscle is intermixed and surrounded by fibroelastic connective tissue. Covered by oral mucosa (see II. A. 1. above) that is tightly adherent to the muscular core. Dorsal surface epithelium is moderately keratinized with projections = **papillae**.

   a) Filiform papillae. Most abundant. Stratified squamous moderately keratinized epithelium projects like a 'dunce hat' with a core of lamina propria. Papillae help produce a rought surface to hold food.

   b) Fungiform papillae. Abundant and scattered among the filiform. Mushroom shape (smooth upper surface) with primary and secondary cores of lamina propria. Contain **taste buds**.

   c) Circumvallate papillae. Least numerous. Exist in a V-shaped organization at **sulcus limitans**. Deep invagination or moat surrounds each papilla. Taste buds occur along the furrow. Serous glands of von Ebner open into the depth of the moat; their secretions keep moat clean and dissolve material to be tasted. Only 10-12 circumvallate papillae.

   d) Taste buds. Composed of neuroepithelial and basal cells (stem cells). **Neuroepithelial cells** have microvilli projecting into a taste pore at the oral surface. Afferent nerve fibers closely appose the neuroepithelial cells. Material enters taste pore, causes a transduction of stimulation to afferent nerves, and information is transferred to CNS for interpretation. There are 4 basic tastes— sweet, salt, sour and bitter.

7. **Tooth**. Hard and soft portions.

   a) Hard portion consists of enamal and dentin. **Dentin** forms the main root portion of the tooth and is produced by odontoblasts throughout life. **Enamel** is the hardest substance in the body, produced by ameloblasts during development and forms the crown overlying the dentin. Dentin rests in the tooth socket of the alveolar bone and this root portion is covered by cementum. **Cementum** is similar to bone and is synthesized by cementocytes (similar to osteocytes). Collagen fibers of the **periodontal ligament** insert in the cementum and the alveolar bone to anchor the tooth.

   b) Soft portion is the pulp. **Pulp** is contained within a cavity surrounded by dentin = pulp cavity. Pulp consists of blood vessels, nerves and mesenchymal connective tissue. Pulp cavity opens at the root of the tooth as the root canal. Odontoblasts are at the border of the pulp adjacent to the dentin.

   c) Attachments of the tooth include the **periodontal ligament**. Collagen fibers embedded in alveolar bone and cementum forms a ligament suspending the tooth. High rate of turnover of collagen here; thus, any process that impairs protein synthesis or vitamin C deficiency, causes atrophy of the ligament. Then the tooth becomes loose in the socket (eventually the tooth could fall out).

   d) Gingiva. Is the mucous membrane connected to the periosteum of the bone by lamina propria and to the tooth by the epithelial attachment. Consists of oral mucosa and is continuous with the 'gum'.

III. General plan of tubular GI organs. There are several **layers** to the wall that are modified to accomodate the function of a particular part of the GI tract. The four main layers to the wall are:

A. **Mucosa**. Lines the lumen. Consists of the epithelium, lamina propria and muscularis mucosa.
  1. Epithelium is wet (moistened by mucus from glands and goblet cells). Epithelial type varies with the function of each segment of the GI tract, but serves for protection, secretion and absorption.
  2. Lamina propria. Ordinary loose connective tissue. Highly cellular with an infiltration of lymphocytes, plasma cells and other blood-derived cells. Is rich in blood and lymph capillaries for absorption of nutrients and dispersal of defense reactions.
  3. Muscularis mucosa. Thin layers of smooth muscle at boundary of mucosa and submucosa. Contraction moves mucosa to facilitate blood flow, mixing of contents of lumen, transport of food materials, and secretion of glands.

B. **Submucosa**. Moderately dense connective tissue layer rich in elastic fibers. Contains larger blood vessels that distribute to the mucosa. Allows folding of the mucosa during empty states, but expansion during filling and/or passage of food. Contains mucous glands in the esophagus and duodenum.

C. **Muscularis externa**. Consists of 2 layers of smooth muscle, inner circular layer and outer longitudinal layer (third layer added in stomach). Maintains integrity of the wall and promotes mixing and movement of food contents via **peristalsis**. **NOTE**: skeletal muscle, under voluntary control, is present only in the pharynx and upper esophagus.

D. **Adventitia/serosa**. Outermost layer. Adventitia = fibrous connective tissue containing fat and larger blood vessels and nerves. Serosa = present where GI tract is covered by a mesentery; this moist membrane provides frictionless movement of organs against each other.

E. General plan of **innervation**. Sympathetic, parasympathetic and sensory nerves supply the GI tract. Two nerve (ganglionated) plexuses in wall: **Meissner's** or submucosal plexus (in submucosa) and **Auerbach's** or myenteric plexus (in muscularis externa).
  1. Submucosal plexus. Contents = postganglionic parasympathetic neurons, sympathetic postganglionic axons and sensory fibers. Supply innervation to both blood vessels and smooth muscle of the muscularis mucosa and lamina propria.
  2. Myenteric plexus. Contents = same as submucosal plexus. Located between inner and outer layers of muscle and innervates the muscularis externa.
  3. Function. In general, **sympathetic impulses** tend to inhibit muscle contraction, tone and glandular secretion (except in sphincters where they stimulate contraction). In general, **parasympathetic impulses** augment muscle contraction, tone and glandular secretion (except in sphincters where they are inhibitory). Sensory or visceral afferents transmit pain, feeling of distension and certain reflexes.

IV. Lower GI system.

A. **Esophagus**. Moistens and conveys masticated food to the stomach. Strong muscular tube with longitudinal folds (due to contraction of muscularis mucosa and externa) to provide distensibility during passage of food bolus.
  1. Mucosa. Lined by stratified squamous non-keratinized epithelium (suited for protection from mechanical stresses). Moistened by mucosal mucous glands in the upper and lower parts of esophagus and submucosal mucous glands in the middle 1/3.

2. Unremarkable lamina propria and submucosa.
3. Muscularis externa. Mixture of smooth and skeletal muscle.
    a) Upper 1/3 contains abundant skeletal muscle, middle 1/3 contains a mixture of smooth and skeletal muscle, lower 1/3 contains smooth muscle.
    b) Physiological sphincters—no muscle thickening, but increased tone—at junctions of esophagus with pharnyx and with stomach.
4. Adventitia. Present, except at the lowest part of esophagus, where it pierces the diaphragm, has a serosa.

B. **Stomach**. Reservoir for ingested food and secretes digestive enzymes and HCl to initiate digestion. Some absorption occurs here (e.g., water, alcohol and salts). Temporary folds of the mucosa and submucosa = **rugae**; gives the appearance of a rough inner surface and allows for marked expansion of the organ.
   1. Mucosa. Surface epithelium, abrupt change from esophagus. Consists of simple columnar epithelium (mucous secreting sheet). Mucus produced by these cells is a glycoprotein that protects the acidified content of the lumen from eroding the gastric mucosa. Invaginations of the surface epithelium form simple tubular glands; numerous indentations of the surface are evident = openings of the **gastric glands** as **gastric pits** or **foveolae**. Three glands are described based on location: cardiac glands, fundic (gastric) glands, and pyloric glands.
      a) **Cardiac glands**. Near esophageal opening; short duct (neck) and highly branched secretory portion. Purely mucous-secreting cells.
      b) **Fundic glands**. Located in body of stomach. Long straight gastric glands; epithelium contains 4 cells types.
         (1) Neck mucous cells. In the upper part (neck) of the gland. Produce mucus and regenerate surface epithelium.
         (2) Parietal cells. Mainly in the middle part of the gland. Eosinophilic. Produce **HCl** through an elaborate intracellular canalicular system, abundant mitochondria and sER. Also produce **gastric intrinsic factor** (important for absorption of vitamin B12 in ileum). **NOTE**: B12 is important for erythropoiesis; lack of B12 results in pernicious anemia.
         (3) Chief cells. Mainly in the lower part of the gland. Basophilic. Produce enzymes, primarily **pepsinogen.** In acid pH, pepsinogen is converted to pepsin which is the active proteolytic enzyme (starts protein digestion).
         (4) Enterochromaffin cells (enteroendocrine cells). Located at base of glands and at the basement membrane; secrete into the vasculature, <u>not</u> into the lumen of gland. Secrete hormones and other bioactive substances: e.g., **serotonin** (vasomotor and increases gut motility) and **gastrin** (stimulates secretion of HCl by parietal cells).
      c) **Pyloric glands**. Near transition to duodenum; long duct (neck) with wide opening; consist of mainly mucous-secreting cells, but some enteroendocrine cells. Hormonesproduced here are **gastrin** and **somatostatin** (acts as local inhibitor of other endocrine cells). Alkaline mucus begins to neutralize the acid chyme as it enters the duodenum.
      d) Regeneration of epithelium. There is a constant turnover of surface cells; replaced by renewal from stem cells of neck mucous cells. Complete turnover of lining cells occurs about every 4 days. **NOTE**: drugs that interfere with mitosis (chemotherapy) have severe consequences for the gut lining.

e) Lamina propria. Very cellular with all of the cells found in loose connective tissue or loose lymphatic tissue. Many cells of the immune system (e.g., plasma cells and lymphocytes).

f) Muscularis mucosa. May have 3 layers.

2. Submucosa. Uusual structure and compositiion.

3. Muscularis externa. 3 layers of smooth muscle; inner oblique, middle circular and outer longitudinal. Middle layer thickens as pyloric sphincter.

4. Serosa. Usual structure and compostion.

5. Control of gastric motility and secretion. Neural (see III. E. above) and hormonal. **Gastrin** secreted by pylorus stimulates pepsinogen and HCl secretion and increased motility. **Secretin** produced by duodenal mucosa decreases motility and secretion of the stomach.

C. **Small intestine**. Continues digestion of food, begins absorption of constituents into blood and lymph vessels and forwards chyme along the tube.

1. General features. Wall has specializations to increase the surface area available for absorption and increase the area available for glands. Specializations include increased **length** (6-8 meters), permanent folds of the mucosa + submucosa (**plicae circularis**) project into the lumen (increase surface area 2-3 times), projections of the mucosa into the lumen (**villi**)(increase surface area 8 times), and projections from the apical surface of absorptive cells (**microvilli**) (increase surface area 20-40 times).

2. Mucosa

a) Epithelium. Simple columnar epithelium with primarily **absorptive cells**, but also abundant mucous-secreting **goblet cells**. Mucosal projections into the lumen = villi. Invaginations of mucosa between villi = intestinal mucosal glands (**crypts of Lieberkühn**).

b) Lamina propria. Prototype of loose connective tissue = loose lymphatic tissue. Many lymphocytes and plasma cells and other cells of connective tissue. Rich in blood capillaries and contains lymph capillaries (**lacteals**). Many nerve fibers here.

c) Muscularis mucosa. Normal constituent of smooth muscle.

3. Submucosa. Moderately dense irregular connective tissue which forms the core of the plicae circularis. **NOTE**: in the duodenum there are characteristic submucosal mucous-secreting glands = Brunner's glands. The submucosa also has many large blood vessels, and Meissner's submucous plexus and ganglia.

4. Muscularis externa. 2 layers of smooth muscle, inner circular and outer longitudinal. Innervated by Auerbach's plexus and ganglia between muscle layers.

5. Serosa. Consists of mesothelium (simple squamous epithelium) resting on a thin layer of loose connective tissue = **visceral peritoneum**.

6. Detail of the mucosa of the small intestine.

a) Mucosal projections = villi and mucosal invaginations between villi = intestinal glands (crypts of Lieberkühn).

b) Epithelial lining of the mucosa (of villi and glands) is simple columnar.

c) Epithelial cell types covering villi.

(1) **Absorptive or principal cells**. Possess many microvilli at their apical surface. Lateral borders are attached by junctional complexes. Absorb materials by 3 means: active transport, pinocytosis and diffusion.

(2) **Goblet cells**. Unicellular mucous glands. Secretion lubricates and neutralizes acid chyme from stomach.

(3) **Enteroendocrine** (enterochromaffin, argentaffin) **cells**. Secrete hormones, e.g., **secretin** (stimulates non-enzyme alkaline secretions from pancreas to help neutralize chyme; inhibits stomach), **cholecystokinin** (stimulates digestive enzyme secretions from pancreas and contraction of gall bladder), **motilin** (increases gut motility), **somatostatin** (inhibit local endocrine cells), **5-hydroxytryptamine** (vasomotor and increases gut motility), **gastric inhibitory peptide** (inhibition of gastric acid secretion).

   d) Epithelial cell types lining crypts.

      (1) **Undifferentiated (stem) cells**. Source for renewal of surface epithelium. Normal structure and function of epithelium of villi depends on balance between cell loss at surface and cell proliferation in crypts. Interference with mitotic activity of undifferentiated cells disrupts function of small intestine. Turnover time of epithelium is about every 4 days.

      (2) **Paneth cells**. Produce lysozyme which is a proteolytic and bacteriolytic enzyme to protect flora of small intestine.

      (3) **Goblet cells**. Have usual structure and function.

      (4) **Enteroendocrine** (enterochromaffin, argentaffin) **cells**. Secrete hormones (see IV.C.6. above).

  7. Other cell types. **M cells** - special epithelial cells overlying lymph nodules, especially Peyer's patches of ileum. Endocytose antigens and present them to underlying lymphoid cells to initiate an immune response.

  8. Distinguishing characteristics of the 3 parts of the small intestine:

    a) Duodenum. Brunner's glands (submucosal), medium-sized villi, high plicae circularis.

    b) Jejunum. Tallest & most slender villi, very high & slender plicae circularis.

    c) Ileum. Low & broad villi, low plicae circularis, aggregates of lymph nodules in lamina propria = Peyer's patches.

D. **Large intestine** (colon). General functions include absorption of water and vitamins derived from the intestinal flora and secretion of mucus to lubricate the contents of the lumen.

  1. Mucosa. Is smooth because it lacks villi and plicae circularis. Epithelial lining is simple columnar with mainly goblet cells, but with some absorptive cells, undifferentiated stem cells and enteroendocrine cells. There are no Paneth cells. Crypts of Lieberkühn are longer than in small intestine.

    a) Lamina propria. Very cellular with many cells of the immune system and solitary lymph nodules. Plasma cells secrete IgGs which helps control the luminal flora.

    b) Muscularis mucosa. Usual smooth muscle.

  2. Submucosa. Usual type but with some adipose tissue and solitary lymph nodules.

  3. Muscularis externa. Inner circular and outer longitudinal smooth muscle layers. Longitudinal muscle organized as 3 bands (rather than a single sheet). Bands are called **taenia coli**.

  4. Serosa. Usual type. Part of the colon is retroperitoneal, thus part of it would have an adventitia.

  5. Innervation. Like that of small intestine.

E. **Vermiform appendix**. Similar structure as the colon, but without taenia coli. Usually there is abundant lymphoid tissue in the lamina propria and submucosa, including lymph nodules.

F. **Rectum and anal canal.**

1. General features. Rectum extends from the upper part of the 3rd sacral vertebra to the pelvic diaphragm. The anal canal extends below the pelvic diaphragm.
2. Mucosa of rectum. Like that of the colon.
3. Mucosa of the anal canal. The epithelium changes from simple columnar to stratified cuboidal or stratified squamous nonkeratinized.
   a) Muscularis mucosa disappears and lamina propria and submucosa merge. Plexus of small veins here = **hemorrhoidal plexus**. Circumanal sweat glands here = apocrine sweat glands.
4. Muscularis externa. In the rectum it has the usual appearance. In the anal canal the inner circular layer thickens to form internal anal sphincter (involuntary sphincter). Beyond the internal sphincter, skeletal muscle accumulates to surround the anal orifice as the external anal sphincter (voluntary).
5. Adventitia. Forms the outer layer.

V. Glandular derivatives of the intestine.

A. Pancreas. Has 2 components, exocrine and endocrine (**islets of Langerhans**).
   1. **Exocrine**. Serous secreting gland that produces digestive enzymes. Duct system drains into duodenum.
      a) Acinar (gland) cells. Cytological prototype of cells producing a protein for export. Produce proteases, lipases, amylases, nucleases, trypsinogen, chymotrypsinogen and others. Cells are organized as acini around a duct system.
      b) Duct cells. Characteristic centroacinar cells which are the initial cells of the intercalated ducts and located at the center of the acini. Duct cells add bicarbonate, water and various salts to the secretion.
      c) Control of secretion is both hormonal and neural.
         (1) Hormones. **Cholecystokinin**, from duodenum, stimulates secretory cells. **Secretin**, from duodenum, stimulates the duct system.
         (2) Neural. Parasympathetic vagal system stimulates the secretory units.
   2. **Endocrine**. Produces mainly the hormones **insulin** and **glucagon**. Both are important for normal carbohydrate metabolism (see ENDOCIRNE SYSTEM for more details).

B. **Liver.**
   1. General functions. Considered both an **endocrine** gland (produces glycose and proteins that are released into capillaries) and an **exocrine** gland (produces bile, carried by a duct system). Synthesis of glycogen, bile, serum proteins (e.g., albumin, fibrinogen, globulins and prothrombin) and serum lipoproteins; detoxification of drugs (alcohol) and reduction of some steroid hormones; filtration of blood (phagocytosis of particulate matter by macrophages). **NOTE**: in general, the liver is concerned with modification of the composition of blood.
   2. General anatomy. Covered by a mesothelium (visceral peritoneum). Underlying the mesothelium is a connective tissue capsule (**Glisson's capsule**). From the capsule, connective tissue trabeculae enter the liver parenchyma to provide structural support. Fine supportive stroma of the liver is **reticular connective tissue**. On the inferior surface is the **hilus** where blood vessels, lymph vessels, nerves and hepatic ducts enter/exit the liver.
   3. **Blood supply**. Liver receives 25% of cardiac output. Liver structure is organized around its blood supply.
      a) Receives blood from 2 sources. Venous blood (70% of flow) from the GI tract that is nutrient-rich, via the **portal vein**; systemic arterial blood (30% of flow) that is

oxygen-rich, via **hepatic artery**. Branches of these vessels enter the hilus and distribute with the trabeculae to areas at the peripheral edge of plates of hepatocytes called **portal canals** (portal triad, portal area, portal radicle).

b) Smallest terminal branches of the <u>artery and vein</u> at the portal canal simultaneously drain into sinusoids among hepatocytes; the venous and arterial blood are mixed in the sinusoids around hepatocytes. Mixed blood flows toward a **central vein** in the center of an hepatic lobule.

c) Hepatic veins drain the central veins and carry venous return from the liver to the inferior vena cava and then to the heart.

4. Principle cells = **hepatocytes**. They are parenchymal cells that are organized as a 3-D system of plates within a vascular framework made of sinusoids. **NOTE**: bile formed by hepatocytes is transported through membrane-lined spaces (bile canaliculi) between hepatocytes to bile ducts; the flow of bile is opposite to the flow of blood.

5. Structural plan of the liver microanatomy. 3 ways to define the liver microanatomy are to consider a classic liver lobule, a liver acinus and a portal lobule.

a) **Classic liver lobule** or hepatic lobule. Consists of a 3-D system of plates of hepatocytes radiating from a central vein. Peripheral boundary of the lobule is 3 **portal triads** (portal canals). Each portal triad contains a branch of an hepatic artery, portal vein and bile duct, as well as a lymph vessel and nerve fibers surrounded as a unit in connective tissue. Thousands of lobules constitute the liver.

(1) Between the plates of hepatocytes are blood-containing sinusoids that drain to the central vein. Sinusoids are discontinuous capillaries lined by endothelial cells **and von Kupffer cells** (macrophages, phagocytose debris in the sinusoidal blood). Material can exit the lumen of the sinusoid to enter a space between the endothelial cell and hepatocyte = **space of Disse.** The space of Disse contains microvilli of hepatocytes that come into contact with soluable elements of the blood. Endocrine secretory products of the hepatocyte are released into the space of Disse and eventually into the sinusoid.

(2) Bile formation and secretion. Bile is the exocrine secretion of the liver hepatocyte and intimately involves the sER. Consists of water, electrolytes, bile acids, bilirubin and cholesterol. Is secreted into a bile **canaliculus** that begins as a blind-ending membrane pouch near the central vein. Canaliculus is a space between the plasma membranes of 2 adjacent hepatocytes that is sealed by tight junctions. Bile drains toward the portal triad and into small ductules lined by squamous cells (canal of Herring) which then empty into cuboidal-cell lined hepatic ducts in the portal triad. Eventually these intrahepatic ducts drain to extrahepatic ducts and then to the gall bladder.

(3) Liver zonation and activity. Activity of a hepatocyte depends on its location within the lobule. There exists a zonation of activity. Cells nearest the periphery of the lobule (closest to the portal triad) (Zone 1) receive the most nutrient- and oxygen-rich blood. Cells around the central vein (Zone 3) receive the most nutrient- and oxygen-poor blood. Cells midway (Zone 2) receive intermediate quality blood. Zone 1 cells 'see' and take up glucose & store glycogen first; they also 'see' and react to toxins and drugs first. Pathological effects are reflected in the zonation.

b) **Liver acinus** (functional unit of the liver). A second way to view liver microanatomy. Liver acinus is built around the terminal branches of the arteries, veins and secretory (bile) duct system. Diamond-shaped structure with a central vein at each end and with terminal branches of the hepatic artery, portal vein and bile ducts at its middle. This unit reflects metabolic and pathologic changes described above for the zonation pattern.

c) **Portal lobule**. Is a third unit to describe the liver microanatomy. It defines the liver lobule organized around the exocrine unit, i.e., the branching pattern of the biliary tree (duct system). A portal triad would be at its center. This is the way the organization of exocrine glands is viewed.

6. Regeneration. The regenerative capacity of hepatocytes is remarkably high. Injury or loss of hepatocytes triggers mitosis in the remaining cells. Normal structural relationships are maintained except in repeated regenerative cycles in which normal architecture becomes distorted, e.g., cirrhosis. In cirrhosis, hepatocytes are progressively replaced with connective tissue and then liver function is reduced (e.g., in alcoholism).

C. **Gall bladder.**

1. General. Functions to store bile produced by liver, concentrates it and reduces its alkalinity.

2. Biliary tree. Bile canaliculus → bile ducts → intrahepatic ducts → common hepatic duct → cystic duct enters the gall bladder.

3. Bladder wall.

a) Mucosa. Epithelium is tall simple columnar epithelium. Cells have many apical microvilli and wide intercellular spaces between cells. Lamina propria merges with submucosa since a muscularis mucosa is lacking. Mucosa thrown into folds (Rokitansky-Aschoff sinuses) that flatten out as the organ fills and distends.

b) Muscularis externa. Thin layers of smooth muscle irregularly arranged and mixed with elastic fibers. They form sphincters in some areas (sphincter of Oddi).

c) Adventitia. Is present where gall bladder is embedded in the liver; is covered by a serosa elsewhere.

4. Concentration of bile. Epithelial cells reabsorb NaCl from luminal contents by active transport to the intercellular spaces. Then water follows passively due to the concentration gradient and passes into the capillaries of the lamina propria.

5. Control of gall bladder. Discharges bile into duodenum. **Cholecystokinin** stimulates contraction of gall bladder smooth muscle. Innervation is similar to that of the tubular GI organs

6. Stones. Concretions of calcium carbonate and cholesterol. Form when foreign substances appear in the bile or when the relative composition of the bile changes causing precipitation of material. Common in people over 50 years old, especially in overweight females.

URINARY SYSTEM
Components: kidney, ureter, urinary bladder and urethra.

I. **Kidney.**

A. General functions. Include filtration of blood, regulation of composition of blood (ions, pH, water, osmotic pressure), absorption of nutrients, secretion/elimination of wastes = production of urine. Kidney also produces hormones (renin and erythropoietin).

B. General morphology. Covered by a connective tissue capsule. On the medial side is the hilus where arteries, veins and nerves enter/exit. Also at the hilus is an expansion of the ureter into the kidney = **renal pelvis.** The pelvis branches as major and minor calyces. These structures are the first part of the excretory passages and collect urine. Major divisions of the kidney are the cortex, medulla, lobes and lobules.
    1. **Cortex**. Filtration of blood occurs here via renal corpuscles.
    2. **Medulla**. Concentration of filtrate into provisional urine occurs here the tubules and vasa recta.
        a) **Medullary** (renal) **pyramid**. Pyramidal-shaped collection of tubules forming the medulla and whose apex opens into a minor calyx.
        b) **Medullary rays**. Groups of medullary tubules (straight parts of nephron and collecting ducts) that radiate into the overlying cortex.
    3. **Lobe**. Consists of a medullary pyramid plus its overlying cortex.
    4. **Lobule**. Consists of a medullary ray plus surrounding cortical material. A medullary ray is the center of a lobule and an interlobular artery on each side forms the lateral boundaries.

C. Functional unit of kidney = **uriniferous tubule**. Consists of a nephron plus its collecting duct.
    1. Nephron. Comprises a renal corpuscle plus the proximal and distal tubules and thin segment associated with that corpuscle.
    2. Renal corpuscle. Comprises a glomerulus (capillary tuft) plus Bowman's capsule.

D. Blood supply. **NOTE**: it is important to understand the blood supply in order to understand the function of the uriniferous tubule. Profuse blood supply to the kidney.
    1. Renal artery → interlobar artery (between medullary pyramids) → arcuate artery (end artery arching over pyramids at corticomedullary junction) → interlobular artery (in cortex, one boundary of a lobule) → afferent arteriole → glomerulus (capillary tuft of renal corpuscle) → efferent arteriole.
    2. Efferent arteriole feeds different capillary beds depending on the location of its nephro
        a) Efferent arterioles of nephrons in the outer cortex → peritubular capillaries (around proximal & distal tubules; nourish tubules and carry away reabsorbed ions and low molecular weight molecules) → stellate veins → interlobular veins → arcuate veins → interlobar veins → renal vein.
        b) Efferent arterioles of juxtamedullary nephrons → **vasa recta** (capillaries around tubules descending into medulla; part of the counter current exchanger/multiplier system for concentration of urine) → arcuate veins → interlobar veins → renal vein.

E. Placement of nephrons. Nephrons have different blood supplies and functional roles in urine production depending on their placement in cortex.
    1. Nephrons in outer cortex. Blood supply is derived from peritubular capillaries and they mainly function in filtration of blood. Peritubular capillaries carry away reabsorbed ions and low molecular weight molecules to help maintain balance in the interstitium.
    2. Nephrons in deep cortex (juxtamedullary nephrons). Blood supply is derived from vasa recta and they function in producing low volumes of concentrated urine.

F. **Renal corpuscle**. Part of nephron and consists of glomerulus plus Bowman's capsule.
    1. **Glomerulus**. Tuft of capillaries. Receives blood at high pressure from afferent arteriole. Hydrostatic pressure important for filtration. Capillaries are fenestrated, held together

by tight junctions and surrounded by a conspicuous basal lamina. Glomerulus is located within a blind pouch = Bowman's capsule.

2. **Bowman's capsule** consists of two layers.
   a) **Visceral layer** (podocytes) = simple squamous epithelium. Podocytes have pedicels (foot processes) that abutt and surround capillaries of the glomerulus.
   b) **Parietal layer** = simple squamous epithelium. Continuous with the visceral layer and with cuboidal cells of the proximal convoluted tubule.
   c) The space between visceral and parietal layers is the **urinary space** and it is continuous with the lumen of the proximal convoluted tubule.

3. **Vascular pole** of renal corpuscle. Area where afferent/efferent arterioles enter/exit. A specialization occurs here: **juxtaglomerular apparatus = macula densa + juxtaglomerular (JG) cells**. Distal convoluted tubule comes close to afferent arteriole of vascular pole and special cells seen in distal tubule wall = **macula densa.** Specialized, modified smooth muscle cells of the afferent arteriole present near macula densa = **JG cells.**

4. **Urinary pole.** On opposite side of corpuscle from vascular pole. Here parietal layer of Bowman's capsule is continuous with the proximal convoluted tubule.

5. **Glomerular filtration membrane** (barrier). Site of filtration of blood. Consists of glomerular capillary endothelium, podocytes and their basal laminae.
   a) Capillary endothelial cell. Is fenestrated (50-90nm pores) and without diaphragms.
   b) Podocyte. Shaped like an octopus and sits on glomerular capillary. Foot processes (pedicels) of podocytes cover the endothelium, but there are spaces (25 nm) between pedicels = **filtration slit**. Slits are closed by diaphragms (that are thinner than a plasma membrane).
   c) Fused basal laminae of the endothelium and podocytes form the **glomerular basal laminae**. Basal laminae contains type IV collagen which helps form a physical barrier for substances. Negatively charged heparan sulfate here acts as a charge barrier.
   d) Filtration. Capillary hydrostatic pressure is important.
      (1) Formed elements of blood (cells and platelets) are excluded (filtered) at the capillary fenestrae.
      (2) Molecules > 10 nm diameter and high molecular weight proteins, > 70,000, do not readily cross basal laminae. The basal laminae are the main filtration mechanism.
      (3) The role of the diaphragms of filtration slits in filtration is not fully agreed upon. It likely acts as a barrier to very large molecules.
      (4) Molecules of molecular weight < 60,000 cross all barriers and into the urinary space as part of the **glomerular filtrate**.
   e) Mesangial cells. Among the capillaries. May be phagocytic and clean filtered material from the basal laminae.

G. **Nephron**. Includes renal corpuscle, proximal tubule, thin segment and distal tubule.
   1. Glomerular filtrate (provisional urine) enters the urinary space (Bowman's space) which becomes continuous with the lumen of the proximal convoluted tubule (see above, I. F. 2. and I. F. 4.).
   2. **Proximal convoluted tubule**. Tall simple cuboidal cells with prominent microvillus border at the apical surface (luminal) and infoldings basally with many mitochondria. Active endocytosis (of proteins) at the apical part of cell. Reabsorbs 80-85% of

glomerular filtrate. Na$^+$, glucose and amino acids are actively reabsorbed, Cl$^-$ and water follow passively.

3. Straight part of proximal tubule (thick, descending part of the loop of Henle).
4. Straight part of distal tubule (thick, ascending part of the loop of Henle).
5. **Distal convoluted tubule**. Simple cuboidal epithelium—not as tall as cells of proximal convoluted tubule, fewer microvilli and mitochondria. Reabsorbs Na$^+$ (influenced by aldosterone), 'secretes' K$^+$, H$^+$ and NH$_3$$^+$ and thus, is involved in **acid-base balance**.
6. **Thin segment**. Simple squamous epithelium that connects proximal and distal tubules. Permeable to water and is involved in creating the hypertonic urine.

H. **Collecting duct**. Epithelium of distal convoluted tubule become continuous with the simple columnar epithelium of the collecting ducts . These cells have few microvilli, but distinct lateral borders. Cells stain poorly. Responsive to **ADH** (antidiuretic hormone or vasopressin—released from posterior pituitary). In presence of ADH, ducts become permeable and water is reabosorbed to aid in producing hypertonic urine.

I. **Papillary duct (of Bellini)**. Largest collecting ducts that open into the minor calyx creating the area cribrosa (area of holes where ducts have opened into minor calyx).

J. **Juxtaglomerular apparatus**. Two components: macula densa and JG cells.
1. **Macula densa**. As the distal convoluted tubule approaches the vascular pole of the renal corpuscle a dense accumulation of specialized cells appear as a spot in its wall = macula densa. Monitors the fluid (chloride content) in the lumen of the tubule.
2. **JG cells.** Modified smooth muscle cells in the afferent arteriole wall (some in efferent arteriole also). Are secretory and produce a proteolytic hormone = **renin**.
3. Functions in blood pressure regulation. Decrease in blood pressure results in release of renin which converts angiotensinogen to angiotensin I; I is converted to angiotensin II which has two actions:
   a) stimulate contraction of arteriolar smooth muscle which increases blood pressure.
   b) stimulate adrenal cortex release of aldosterone, which leads to sodium reabsorption; water follows sodium to increase blood volume and thus blood pressure.

II. Excretory passages: minor calyx, major calyx, renal pelvis, ureter, bladder and urethra. Lined by transitional epithelium which is specialized to accommodate distension and withstand the hypertonicity of the urine.

A. Calyces and pelvis are lined by transitional epithelium and receive the urine from papillary ducts of Bellini. Lamina propria is usual variety. Muscularis composed of 2 ill-defined layers of smooth muscle helically arranged (inner layer more longitudinal, outer layer more circular). Muscle contraction helps propel urine.

B. Ureter.
1. Mucosa. Lined by transitional epithelium resting on a thin fibroelastic lamina propria.
2. Submucosa is lacking.
3. Muscularis. 2 layers in upper part (inner longitudinal and outer circular; opposite of GI system) and 3 layers in lower part (an outer longitudinal layer is added). Peristaltic contraction propels the urine.
4. Adventitia. Fibroelastic connective tissue containing arteries, veins and nerves.
5. Innervation. Sympathetic, parasympathetic and sensory. **NOTE**: passage of kidney stones is very painful. Sensory fibers are also involved in reflexes.

C. Bladder.
1. Mucosa. Like that of the ureter.
2. Muscularis. 3 ill-defined layers similar to the lower part of the ureter. Circular muscle is arranged around the urethral outlet = **internal bladder sphincter**.
3. Serosa. Covers the upper surface, adventitia covers the rest of the bladder.

D. Urethra.
1. Female. 4-5 cm long. Lined by transitional epithelium at the bladder but progressively changes to stratified squamous near external opening. Mucous glands pouch off the the epithelial lining. A thin lamina propria contains a plexus of veins. Ill-defined layers of smooth muscle. Skeletal muscle at the external orifice = **external urethral sphincter** (voluntary control).
2. Male. 15-20 cm long. 3 parts: **prostatic**, **membranous** and **penile**. Lined by transitional epithelium (prostatic), stratified columnar (membranous) and pseudostratified to stratified squamous (penile). Membranous part traverses the urogenital diaphragm (skeletal muscle) which forms the external bladder sphincter. Mucous-secreting glands of Littre open into penile urethra, as well as the bulbourethral (Cowper's) glands. The latter release a viscous fluid on sexual excitement.

ENDOCRINE SYSTEM

I. Overall function. Maintenance of a physiologic steady state in the face of continous changes; maintain homeostasis. This is done by chemical messengers (hormones) secreted in response to various stimuli and carried by the blood.

II. Endocrine glands.

A. Classical glands. Include pituitary (hypophysis), thyroid, parathyroid, adrenal and pineal glands, islets of Langerhans (in pancreas), and gonads (testis and ovary).

B. Non-classical glands. Include liver, kidney, thymus, placenta, GI tract and undoubtedly others (NOTE: these will not be considered in this section).

C. General characteristics of classical glands.
1. Origin. Epithelial, except posterior pituitary and adrenal medulla which are neural in origin.
2. Glands lack a duct system. Secretion of products is into capillaries.
3. Profuse vascularization. Fenestrated capillaries with diaphragms.
4. Hormones. Are either proteins, glycoproteins, peptides or steroids.
   a) Cellular structure reflects product. Steroid-hormone producing cells have abundant sER and mitochondria. Cells producing peptides or proteins have abundant rER and secretory granules. Cells producing glycoproteins have abundant sER, rER and secretory granules.
   b) Type of hormone reflects mode of storage and release.
      (1) Steroids are not stored, but synthezed and released on demand.
      (2) Many proteins and glycoproteins are produced and stored intracellularly in secretion granules (e.g., cells of pars distalis).
      (3) Some hormones are stored extracellularly (thyroxine in thyroid follicles).
5. Hypothalamus of the brain plays a major role of integration of neural inputs with resultant transferral of secretory signals to pituitary. Hypothalamo-pituitary interaction is

vital for some endocrine glands, but not all (e.g., adrenal medulla and islets of Langerhans are not pituitary dependent).

III. **Pituitary gland** (hypophysis). Attached to base of the brain, at the median eminence of the tuber cinereum, by a stalk (infundibulum).

A. Components and nomenclature. Two lobes: anterior and posterior.
    1. **Anterior lobe = adenohypophysis = pars distalis**, pars tuberalis and pars intermedia (the latter is vestigial in humans). Derived from oral ectoderm (Rathke's pouch).
    2. **Posterior lobe = neurohypophysis = pars nervosa** and infundibular stem. Derived from neural ectoderm of the diencephalon.

B. Blood supply and innervation.
    1. Anterior lobe. **Blood supply.** Superior hypophyseal artery (from the internal carotid artery) forms a primary capillary bed in the median eminence. Venules arise from here and distribute as capillary sinusoids (secondary capillary bed) around cells of anterior lobe, thus constituting the **hypophyseal portal vessels (system)**. The portal vessels carry releasing/inhibitory factors from the median eminence to the cells of the anterior lobe. From the secondary capillary bed arises the venous drainage that carries pituitary hormones to the systemic circulation. **Innervation**: neurons in the area of the median eminence produce releasing and inhibitory factors for anterior lobe pituitary hormones and project their axons to the primary capillary bed where these factors are released.
    2. Posterior lobe. **Blood supply.** Inferior hypophyseal artery (from internal carotid artery) enters the pars nervosa, forms a capillary bed which collects as venules to join the systemic circulation (this is the usual vascular arrangement in organs). However, some venules arising from these capillaries end as capillaries around cells in the anterior lobe, thus shunting blood to the anterior lobe as a portal system (**hypophysioportal system**). This may be an avenue for some factors to get from pars nervosa to influence the cells of pars distalis. **Innervation**: neurosecretory neurons in the hypothalamus (see below) project axons which terminate around, and release hormones into, the fenestrated capillaries of the pars nervosa.

C. Posterior lobe - neurohypophysis, pars nervosa.
    1. Neuron somata in the hypothalamus (supraoptic and paraventricular nuclei) project their axons to the capillary beds in the pars nervosa. Neurons have typical neuronal morphology with the addition of many large, dense-core neurosecretory granules in the cytoplasm (these cells have both glandular and neuronal properties). Axons form the **hypothalamohypophyseal tract.** Supporting elements within the pars nervosa = **pituicytes** (similar to neuroglial cells of the CNS).
    2. Secretion. Hormones are produced in the neuron somata, and transported to the terminals, and released into capillaries in the pars nervosa. Accumulations of hormone along the axons = **Herring bodies**. Hormones are complexed to a carried protein = **neurophysin**.
    3. Hormones. **Arginine vasopressin (antidiuretic hormone, ADH)**. Functions to increase permeability of distal convoluted tubules and collecting ducts of kidney to promote retention of fluid. It also stimulates vasoconstriction to increase blood pressure. **NOTE:** diabetes insipidus results from lack of ADH; patient continuously takes in fluids and continuously produces large amounts of urine. **Oxytocin.** Functions to stimulate contraction of myometrium and myoepithelial cells in lactating mammary gland (promotes milk release).

4. Control of secretion. Vasopressin is released on demand by increases in osmolarity of the blood or by a decrease in blood pressure. Oxytocin is secreted in response to afferent neural stimulation; e.g., suckling stimulus to the nipple and stretch of the vagina or cervix.

D. Anterior lobe - adenohypophysis, pars distalis.
1. Structure. Cords or clusters of cells situated among sinusoidal capillaries. There is profuse vascularization, but little connective tissue. All hormonal secretions are proteins or glycoproteins and are released directly to the capillary system.
2. Cells. 2 types.
   a) Chromophils. 50% of cells and stain well. 2 types of chromophils:
      (1) Acidophils (40%) and include **somatotropes** (produce growth hormone - GH) and **lactotropes** (prolactin - PRL).
      (2) Basophils (10%) and include **thyrotropes** (produce thyroid stimulating hormone - TSH), **corticotropes** (produce adrenocorticotrophic hormone - ACTH) and **gonadotropes** (produce follicle stimulating hormone - FSH - and luteinizing hormone - LH).
   b) Chromophobes. 50% of cells and do not stain well. Are degranulated acidophils and basophils.
3. Hormones and effects.
   a) GH. Stimulates production of insulin-like growth factor (somatomedin) by liver and enhances growth at epiphyseal plates, general body growth, protein synthesis and carbohydrate utilization.
   b) PRL. Promotes mammary gland development during pregnancy and milk synthesis/secretion.
   c) TSH. Promotes structural and functional (synthesis and release of thyroid hormones) aspects of thyroid gland.
   d) ACTH. Stimulates growth and secretion (corticosteroids) of adrenal cortex.
   e) FSH. Stimulates growth and maturation of ovarian follicles. Stimulates androgen binding and thus, spermatogenesis.
   f) LH. Stimulates ovulation and formation of the corpus luteum. Stimulates androgen secretion and thus, spermatogenesis.
4. Control of secretion of cells in anterior lobe. Most hormones have a releasing and/or inhibitory factor produced in hypothalamic neurons that govern their secretion. Also, most are influenced by feedback from the products produced by target organs of the anterior pituitary hormone (e.g., thyroxine from thyroid gland has a negative feedback to hypothalamus and anterior pituitary to reduce production of TSH-releasing factor and TSH, respectively). There is not a negative feedback on acidophils.

E. Pars tuberalis. Wrapped around the infundibular stalk. Site of primary capillary plexus. Similar structure and function to pars distalis.

F. Pars intermedia. Cells (similar to basophils) form vesicles filled with colloid. These are remnants of Rathke's pouch. Function in human is questionable.

IV. **Thyroid gland**.

A. Structure. Two lateral lobes connected by an isthmus at the level of the 2nd and 3rd tracheal cartilages. Enveloped by a capsule and has a fine connective tissue stroma. Well vascularized.

B.   Parenchyma. 2 cell types, follicular cells and parafollicular cells.
1.   **Follicular cells**. Are simple cuboidal (normal) to simple columnar (hyperactive) to simple squamous (inactive) epithelial cells. They form spherical, hollow balls of cells = **follicles**. Follicles are filled with colloid material that consists of **thyroglobulin** (a binding glycoprotein for thyroid hormone) + thyroid hormone (**thyroxine**). Thus, thyroxine is stored extracellularly. Follicular cells contain appropriate organelles for synthesizing protein and glycoprotein. Cells also take up iodine from the vasculature and transport it to colloid where it is added to tyrosines (which are attached to thyroglobulin). For release, thyroglobulin with attached thyroxine is taken up by pinocytosis, is broken down by lysosomes, and thyroxine released into capillaries.
2.   **Parafollicular cells**. Occur within wall of follicle (inside the basement membrane) or as small clusters outside the follicle. Produce the hormone, **calcitonin.**

C.   Function of thyroxine. Increase basal metabolic rate, increase oxidative metabolism, stimulate maturational processes and increase CNS activity. Control of secretion is by TSH from the pars distalis.

D.   Function of calcitonin. Lowers blood calcium levels and decreases bone resorption. Control of secretion is by negative feedback with plasma calcium levels.

V.   **Parathyroid gland**.

A.   Structure. Very small glands embedded in the thyroid gland, but separated by a capsule. Adipocytes are common in the glands, especially with increasing age.

B.   Parenchyma. 2 cell types—chief and oxyphil.
1.   **Chief cells** are most numerous and produce **parathyroid hormone (PTH)**. PTH increases plasma calcium (acts the opposite of calcitonin). Increases bone resorption (by osteoclasts) to increase calcium, stimulates absorption of calcium in the small intestine and reduces renal excretion of calcium. Control of secretion is via negative feedback with plasma calcium levels.
2.   **Oxyphil cells**. Less numerous, but larger than chief cells. Acidophilic cells that resemble parietal cells of the stomach. Function in humans is unknown.

VI.   **Adrenal gland**.

A.   Structure. 2 parts—cortex and medulla with divergent origins. Cortex develops from mesoderm, whereas medulla develops from neural crest (is a modified sympathetic ganglion). Surrounded by a distinct capsule, but there is little connective tissue stroma. Profuse vascularization and capillaries are sinusoidal.

B.   Blood supply. Subcapsular arterioles give rise to sinusoids that:
1.   Supply the cortex first and then drain to medullary veins (these carry corticosteroids to the medulla) or
2.   Supply directly the medulla (so the medulla does not get nutrient-deficient blood) and then collect as medullary veins.

C.   Innervation. Medulla is a modified ganglion. Its cells receive sympathetic preganglionic axons and synapses. In this way sympathetic stimuli can release catecholamines.

D.   **Cortex**. Cells are arranged in zones.
1.   **Zona glomerulosa**. Outermost. Cells arranged as arches surrounded by sinusoids. Cells contain lipid droplets, abundant mitochondria and sER (characteristics of steroid

hormone producing cells). Produce the mineralocorticoid **aldosterone**. Stimulated by angiotensin II.

2. **Zona fasciculata**. Middle layer, forms bulk of cortex. Cells arranged as straight cords, 2 cells thick and separated by sinusoids. Ultrastructure like that of cells in zona glomerulosa, but more 'spongy, foamy' appearing because of lipid content. Produce the glucocorticoids **cortisol** and **corticosterone**. Stimulated by ACTH.

3. **Zona reticularis**. Innermost layer. Cells arranged as a network among sinusoids. Cells are smaller, contain less lipid, but more lipofuscin than those in other layers. Produce the sex steroid **dehydroepiandrosterone** (a weak androgen).

4. Hormonal actions. Aldosterone promotes sodium retention in the kidney. Glucocorticoids promote carbohydrate production form proteins and increase blood glucose levels; also alleviates inflammation and are immunosuppressive. Dehydroepiandrosterone usually has little effect, but overproduction can cause masculinization in females.

E. **Medulla**. Cells are arranged in clusters among sinusoids.
   1. Cell types include chromaffin cells and sympathetic postganglionic neurons.
      a) **Chromaffin cells**. Resemble autonomic neurons, but without axons. They contain abundant rER, Golgi, mitochondria, and neurosecretory, dense-core vesicles. The secretory granules contain the catecholamines **epinephrine** or **norepinephrine.** About 80% of the secretion is epinephrine.
      b) Sympathetic postganglionic neurons. Are few and are displaced from sympathetic ganglia. Their axons innervate the vasculature in the medulla.
   2. Hormonal effects. Epinephrine prepares us for the 'fight or flight' reaction. Increased glycogenolysis, increased cardiac output, increased oxygen consumption, and increased perfusion of skeletal muscle. Norepinephrine increases vasoconstriction and thus blood pressure.
   3. Control of secretion. Via stimulation by sympathetic preganglionic neurons and synaptic contacts. Hypoglycemia can stimulate epinephrine secretion.
   4. Glucocorticoids from the cortex are important in maintenance of the medullary cells as endocrine cells (rather than neurons) and for the production of epinephrine from norepinephrine.

VII. **Islets of Langerhans** (endocrine pancreas).

A. Structure. Clusters of cells situated in the bulk of the exocrine pancreas, but separated by a delicate connective tissue capsule. Profusely vascularized by fenestrated capillaries.

B. Cell types.
   1. **A or alpha cells**. Are peripherlly located and form about 15% of the islet cell population. Produce the hormone **glucagon.**
   2. **B or beta cells**. Are more centrally located and form about 70% of the islet cell population. Produce the hormone **insulin**.
   3. **D or delta cells**. Peripherally located and form about 10% of the islet cell population. Produce the hormone **somatostatin**.
   4. Other cell types (C, F and PP cells) form about 5% of the cell population. Even though they secrete various hormones, their functional roles in the islet presently are not clear.

C. Hormonal effects. Glucagon increases blood sugar by promoting glycogenolysis in the liver (stimulated by low plasma glucose). Insulin decreases blood sugar by promoting glucose transport into muscle and adipocytes and glycogen synthesis in the liver (stimulated by rising blood glucose levels). Somatostatin inhibits the release of insulin and glucagon. **NOTE**: diabetes mellitus results from a deficiency of insulin or beta cells (expressed as hyperglycemia and as glycosuria—increased glucose in the urine).

D. Innervation. Some cells are innervated by sympathetic and parasympthetic nerves (sympathetic stimulation decreases insulin secretion in animals).

VIII. **Pineal gland**.

A. Structure. Attached by the pineal stalk to the dorsal (posterior) aspect of the diencephalon and 3rd ventricle. 5-8mm long by 3-5mm wide. The stalk contains nerve fibers that enter the gland. The pineal is surrounded by a capsule that is continuous with the pia/arachnoid meninges.

B. Cell types. Pinealocytes (chief cells) and glial cells.
    1. **Pinealocyte**. Large, basophilic cells that have processes ending near capillaries. Processes resemble axons and contain synaptic vesicles. Pinealocytes are connected by gap junctions.
    2. **Glial cells**. Resemble astrocytes of the CNS and appear to play a supportive role for pinealocytes.

C. Hormones. Pinealocytes contain **serotonin** and **melatonin** (serotonin is a precursor to melatonin). Melatonin has antigonadotrophic effects. Melatonin production varies with light cycles and is responsive to visual inputs (mainly produced when no light enters the visual system—night). Target of melatonin is hypothalamic cells that produce gonadotropin-releasing hormone (GnRH) and gonadotrophs. Prevents precocious gonadal development and function.

D. Innervation. By sympathetic nerves. Information from the retina travels to preganglionic neurons in thoracic spinal cord then to postganglionic neurons in the superior cervical ganglion. Postganglionic sympathetic fibers synapse on pinealocytes. Thus, visual inputs related to light and dark cycles get to pineal gland.

E. Calcified inclusions = **corpora aranacea** (brain sand). Are frequently present in the pineal. **NOTE**: they increase with age and are used as a radiological landmark.

FEMALE REPRODUCTIVE SYSTEM

I. General function. Perpetuation of the species. Production of germ cells (ova)(stimulated by FSH), transport and fertilization of ova, secretion of hormones to prepare for implantation (FSH and estrogen, followed by LH) and maintenance of implantation (LH, HCG, estrogen and progesterone), forcible expulsion of the fetus (stimulated by oxytocin) and provides nutrition for the newborn by lactation (stimulated by prolactin and oxytocin). The system undergoes 3 phases: **infantile** (birth to puberty, 12-14th year), **functional/reproductive** (12-14th year to 45-55th year) and **involutional** (from 45-55th year to death; menopause designates ovarian involution). Menarche = time of first menses (menstruation; shedding of the uterine mucosa if implantation does not occur). From menarche to menopause the system undergoes cyclic changes governed by hormonal relationships involving the brain, endocrine system and reproductive organs.

II. Components.

A. Primary sex organs. **Ovaries** have dual functions; exocrine (produce oocytes) and endocrine (secrete the hormones estrogen and progesterone).

B. Secondary sex organs. Uterine tubes (oviducts, fallopian tubes), uterus, vagina and external genitalia.

III. Ovary.

A. Immature ovary. Covered by a simple cuboidal epithelium (germinal epithelium; this is a misnomer in that this epithelium does not give rise to oocytes).

1. Cortex. At birth, contains about 200,000 primordial follicles/ovary. **Primordial follicle** = primary oocyte surrounded by a single layer of squamous follicular cells. Primary oocytes are arrested in prophase of first meiotic division. Growing follicles and corpora lutea are absent. Stroma is loose connective tissue.

2. Medulla. Contains blood vessels, nerves and loose connective tissue.

B. Mature (functional) ovary. Dense connective tissue layer (**tunica albuginea**) is evident underlying the germinal epithelium.

1. Medulla. Contains blood vessels, nerves and loose connective tissue.

2. Cortex. Contains many follicles in different maturational phases.

a) **Primordial follicle**. Primary oocyte surrounded by a single layer of squamous follicular (granulosa) cells. With each cycle, under stimulation by FSH, primordial follicles commence to grow to form primary follicles.

b) **Primary (growing) follicle**. Follicular cells become cuboidal, multilayered, and a basement membrane is produced adjacent to stroma. Oocyte enlarges and a **zona pellicuda** is produced by both the oocyte and granulosa (follicular) cells. Oocyte and granulosa cells communicate, via gap junctions, across the zona pellucida. Under continued stimulation by FSH the primary follicle advances to a secondary follicle.

c) **Secondary (antral) follicle**. Spaces begin to form among the granulosa cells = beginning of **antrum**. Spaces contain **liquor folliculi** containing estrogen, progesterone and androgen. Eventually the spaces coalesce as a single antrum. The oocyte is pushed to one side of the follicle in a mound of granulosa cells = **cumulus oophorus**. The stroma immediately around the follicle organizes into concentric layers = **theca interna** and **theca externa**. Theca externa remains as stroma, but theca interna cells becomes steroidogenic (they produce androstenedione and testosterone which are converted to estrogen by the granulosa cells). With continued maturation and stimulation by FSH, the secondary follicle becomes a Graafian follicle.

d) **Graafian (mature) follicle**. Fully mature follicle that exists 1-2 days before ovulation. It moves from deep in the cortex to the sufacge at the site where ovulation will occur (**stigma**). Oocyte is eccentrically located in the follicle and surrounded by the cumulus oophorus. Granulosa cells immediately around and attached to the zona pellicida form the **corona radiata** and will be expelled with the oocyte along with liquor folliculi. Just prior to ovulation, the first meiotic division is completed and the primary oocyte becomes a secondary ooctye (it is arrested at metaphase of the second meiotic division until fertilization occurs).

3. Ovulation. About midcycle there is an abrupt surge in LH secretion which promotes ovulation. The stigma ruptures, the secondary oocyte with its corona radiata is released into the peritoneal cavity and is swept into the infundibulum of the oviduct. If not fertilized within 24 hours the secondary oocyte degenerates. If fertilization occurs, the secondary oocyte completes the second meiotic division.

a) **Corpus luteum (CL)** formation. The remaining wall of the ruptured follicle collapses, granulosa cells become vascularized and luteinization occurs (transformation of cells into steroidogenic cells). Under influence of LH, theca interna cells become **theca lutein cells** (continue to produce estrogen) and granulosa cells become **granulosa lutein cells** (80% of corpus luteum; produce progesterone). Progesterone prepares uterus for implantation and prevents development of new follicles and subsequent ovulation.
   (1) Life span of CL. If pregnancy does not occur, 10-14 days; if pregnancy occurs, 6 months (continues to produce estrogen and progesterone throughout pregnancy, but at a reduced level after 6 months).
   (2) CL is maintained by human chorionic gonadotrophin (HCG) produced by the placenta. In late pregnancy the CL produces relaxin (relaxes the pubic symphysis, inhibits the myometrium and promotes dilation of the cervix). If (when) the CL degenerates, it becomes a **corpus albicans**.
   b) **Corpus albicans**. Degenerated corpus luteum. Becomes clear staining, fibrotic, and contains active macrophages.
4. **Atretic follicles**. A large number of follicles begin to develop with each cycle, but only one completes development and is ovulated. Others degenerate (undgo atresia) at various stages of development = **atretic follicles**. **NOTE**: of the 400,000 follicles that a young female has at birth, only 400-500 are ovulated, the rest undergo atresia. Atretic follicles are recognized by disorganized and degenerating granulosa cells, hypertrophied and folded basement membrane of granulosa cells, and a degenerating oocyte.

C. Hormonal influences.
   1. Estrogen. Produced by growing follicle and corpus luteum.
   Stimulates development and maintenance of female reproductive tract and mammary glands. Stimulates female secondary sex characteristics. Influences CNS (hormonal feedback and behavior).
   2. Progesterone. Produced by corpus luteum.
   Hormone of pregnancy—in general, acts after estrogen to prepare reproductive tract for pregnancy. Stimulates uterine gland secretion and mammary gland development. Inhibits uterine contractions. Influences CNS (hormonal feedback and behavior).
   3. FSH. Produced by anterior pituitary. Stimulates follicular growth and thus, promotes estrogen secretion.
   4. LH. Produced by anterior pituitary Promotes ovulation, corpus luteum formation, and progesterone secretion by the corpus luteum.

IV. **Uterine tube** (oviduct, fallopian tube).

A. Parts. From uterus distally: intramural, isthmus, ampulla, and infundibulum with fimbria. Infundibulum opens to the peritoneal cavity near the ovary and the finger-like fimbria (along with ciliary action) sweep the expelled ovum into the lumen of the tube.

B. Histology.
   1. **Mucosa**. Simple cuboidal to simple columnar cells; some are ciliated and some are secretory cells. Height of epithelial cells, presence of cilia and secretory activity are all stimulated by estrogen. Some cilia beat toward the tube to move ovum into tube, others beat toward ovary to help move sperm to ovum. Secretory cells produce viscous material to nourish ovum and promote capacitation of sperm. **Lamina propria** is thin,

vascular and without glands. Mucosa is folded with most complex folds in the ampulla; fimbria are inverted folds.

2.   Muscularis. Inner circular and outer longitudinal smooth muscle (like GI tract). Increases in thickness toward the uterus and peristaltic contraction waves to propel the oocyte toward the uterus. Stimulated by estrogen (during the follicular or proliferative phase) and inhibited by progesterone (during the luteal or secretory phase).

3.   Serosa. Tube is covered by mesothelium = visceral peritoneum.

C.   Function. Moves ovum and sperm together to promote fertilization, site of fertilization, movement of fertilized ovum to uterus for implantation. **NOTE:** mucosa is well vascularized and can serve as a site for implantation = <u>ectopic pregnancy</u>.

V.   Uterus.

A.   General function. Receives the fertilized ovum (zygote), provides for its attachment (implantation) and establishes a blood supply for maintenance of the embryo (via the placenta).

B.   Parts. Body (with upper dome-shaped fundus) and cervix (most caudal). Body consists of 3 distinct layers and undergoes cyclic changes during the menstrual cycle. In cervix, layers are less distinct and cyclic changes are not pronounced.

C.   **Uterine body**. Wall consists of 3 layers.

1.   **Endometrium** (mucosa). Epithelium is simple columnar to cuboidal; it invaginates from the surface to form simple tubular uterine glands. Epithelium contains some secretory cells (mainly in glands) and ciliated cells (mainly at surface). Lamina propria is very loose and cellular. Is divided into 2 layers:

a)   **Stratum functionalis**. Luminal layer. Undergoes cyclic changes under the influence of estrogen and is sloughed during menstruation.

b)   **Stratum basalis**. Deep layer abutting myometrium. Does not undergo significant cyclic changes. Regenerates the stratum functionalis after menstruation (during the follicular or proliferative phase).

2.   **Myometrium** (muscularis). Three ill-defined layers of smooth muscle. Undergoes hyperplasia and hypertrophy during pregnancy; postpregnancy, the muscle does not return to its original state, but there remains some uterine enlargement. Is well vascularized.

3.   **Perimetrium** (serosa). Typical peritoneal lining of mesothelium with a small amount of underlying connective tissue. Continuous with lining of uterine tubes and broad ligament. Blood vessels and nerves run in the perimetrium.

D.   Blood supply to endometrium. Is important for menstrual cycle and pregnancy.

1.   **Uterine** and **ovarian arteries** anastomose and give rise to arcuate branches that encircle the uterus. Branches of the **arcuate arteries** penetrate the myometrium as **radial arteries** and terminate at the endo/myometrial junction as **straight** and **spiral (coiled) arteries** which supply the endometrium.

2.   Straight arteries. Mainly supply the stratum basalis.

3.   Spiral arteries. Mainly supply the stratum functionalis. Drain to venous lacunae near the luminal surface which in turn connect into the venous system.

E.   Cyclic changes in endometrium (coincides with the functioning of the ovary—begins at puberty). The endometrium is prepared for implantation; if implantation does not occur there is a partial sloughing of the endometrium (stratum functionalis). Three stages to the cycle: proliferative, secretory and menstrual.

1. **Proliferative** stage (follicular phase—coincides with FSH secretion, growth of ovarian follicles and production of estrogen).

    a) Stratum functionalis is regenerated from the stratum basalis after menstruation. The endometrial glands increase in length and number and the gland lumen is narrow and straight. Spiral arteries lengthen and extend to midlevels of the endometrium. In the later part of this phase, the cells become secretory and the glandular lumen becomes wavy.

    b) Continues from about day 6 to about the 14th day of the cycle = midcycle (day 1 of cycle is first day of menstruation). At midcycle there is an LH surge and ovulation. This leads to:

2. **Secretory** stage (luteal phase—coincides with LH secretion, life of the corpus luteum, and the secretion of progesterone).

    a) Stratum functionalis thickens due to edema of lamina propria and hypertrophy of epithelial cells. Glands become tortuous, lumen is wide and sinuous, and epithelial cells are secretory (secrete glycogen for nutrition of fertilized ovum). Spiral arteries extend close to the luminal surface and venous lakes have many distentions. Stratum functionalis is prepared for implantation.

    b) Continues to about the 28th day of the cycle = end of cycle (day 1 of cycle is first day of menstruation).

3. **Menstrual** stage (ischemic phase—coincides with lack of implantation, decline of corpus luteum function and decline of progesterone and estrogen secretion).

    a) Spiral arteries constrict intermittently, stratum functionalis becomes ischemic, capillaries and venous lakes rupture, epithelium breaks down with a general disruption of the tissue. Fragments of mucosa detach, stroma and vessels are exposed, bleeding occurs and stratum functionalis is sloughed. Average blood loss is about 35 ml.

    b) Continues from day 1 to day 5 of the cycle. Then the proliferative stage begins again.

    c) If implantation occurs then the placenta is formed and produces HCG which rescues the corpus luteum, keeps it viable and secreting progesterone and estrogen. This maintains the endometrium and menstruation does not occur.

F. Changes in the endometrium if fertilization occurs.

    1. Fertilization produces a zygote. Zygote undergoes many mitoses to form ball of cells = **morula.** Morula enters uterine cavity about 3 days after fertilization (about day 17 of the cycle when the endometrium is in the secretory stage). Cavity forms in morula and the **blastocyst** is born.

    2. The outer cell layer of the blastocyst = **trophoblast**. Trophoblast invades the uterine epithelium and gives rise to cytotrophoblast cells and the syncytiotrophoblast. **Syncytiotrophoblast** secretes HCG. HCG rescues the corpus luteum so menstruation does not occur. Trophoblast is the fetal contribution to the placenta and the stratum basalis (decidua basalis) is the maternal contribution to the placenta.

VI. **Uterine cervix.**

A. General. Lowest part of the uterus and connects the uterine body with the vagina.

    1. Mucosa. Epithelium is simple columnar, mucous-secreting cells. Epithelium invaginates forming mucous-secreting glands. Secretion is thinnest (more watery) at midcycle

under estrogen stimulation (most conducive to sperm penetration). Secretion is more viscous after ovulation and during pregnancy (acts as a barrier to sperm and bacteria). **NOTE**: duct of glands often get plugged, which backs up the secretion leading to cyst formation = Nabothian cyst. Mucosa does not slough during the menstrual stage; only cyclic change is the secretory activity of the mucous cells. Mucosa is highly folded forming **plica palmatae.** Lamina propria is dense connective tissue.

   2.   Myometrium. Mixture of smooth muscle, elastic and collagen fibers.

B.   Portio vaginalis. Lower part of cervix protrudes into the vagina as the **portio vaginalis**. The epithelium abruptly changes here from simple columnar to stratified squamous non-keratinized. **NOTE**: cells taken from this area are the basis for the Pap smear.

VII. **Vagina.**

A.   General. Is a muscular tube extending from the cervix to the vestibule. Wall consists of three layers: mucosa, muscularis and adventitia.

   1.   Mucosa. Epithelium is stratified squamous non-keratinized. Cells contain abundant glycogen at midcycle. Lamina propria is dense connective tissue, but becomes loose more peripherally in the wall. Many veins in the loose connective tissue; become engorged during sexual arousal. There are no glands in the vagina; lubrication comes from mucus of the cervix. Many lymphoid cells in the loose connective tissue; aid in patrolling for antigens (especially when the epithelium thins during the luteal phase of the cycle).

   2.   Muscularis. Inner circular and outer longitudinal smooth muscle. There are skeletal muscle fibers at the vaginal opening that form a sphincter.

   3.   Adventitia. Connective tissue rich in elastic fibers.

B.   Epithelial changes with hormones and the menstrual cycle.

   1.   Estrogen stimulates an increase in epithelial thickness and glycogen accumulation (fermentation of glycogen to lactic acid by bacteria is responsible for the acid pH of the vaginal fluid near midcycle—this helps resist infections).

   2.   Drop in estrogen titer leads to a decrease in epithelial thickness in the second half of the cycle due to sloughing of the surface cells. This thinning of the epithelium permits lymphocytes to pass into the vaginal lumen. Glycogen is less abundant, thus pH of the vaginal fluid increases at this time—increased chances for infections at this time.

VIII. **External genitalia.** Includes the clitoris (homologue of the penis), labia minora and majora (folds of skin with a core of connective tissue with elastic fibers), mucous-secreting vestibular glands as well as Bartholin's glands that secrete a milky lubricant during sexual arousal.

IX. **Placenta.**

A.   **Blood-placenta barrier or fetal-maternal barrier**.

   1.   Six layers. Syncytiotrophoblast, cytotrophoblast, basal lamina, connective tissue space, basal lamina and endothelium of capillaries. (During late pregnancy, the cytotrophoblast drops out leaving five layers).

   2.   Prevents mixing of maternal and fetal blood and prevents certain materials from crossing and mixing with fetal blood.

B.   Placenta as an endocrine organ. Secretes **HCG**, **human placental lactogen** (has GH and lactogenic activities), **estrogen** and **progesterone**. These substances are secreted by the syncytiotrophoblasts.

X. **Mammary gland**.

A. Compound tubuloalveolar gland containing 15-20 lobes. Each lobe is drained by a lactiferous duct that exits to the surface of the nipple. Many secretory alveoli (acini) drain into each duct. Lobes are separated by dense connective tissue and adipose tissue (in the adult). The gland is subject to developmental and reproductive changes.

1. **Prepubertal gland**. See only ductal elements surrounded by connective tissue; secretory alveoli are not developed.

2. **Postpubertal, non-lactating gland**. Estrogen stimulates deposition of adipose tissue. Some branching of ducts occurs and secretory alveoli sprout from the ducts. Alveoli consist of simple cuboidal cells.

3. **Adult, non-pregnant, non-lactating gland**. Adipose tissue is prominent. Glands (alveoli) branch from ducts (forming lobules) and are surrounded by loose connective tissue with dense connective tissue between the lobules. Progesterone influences branching of duct system.

4. **Adult, pregnant, lactating gland**. Alveoli are highly branched from ducts and have expanded at the expense of the adipose tissue and loose connective tissue of the lobules. This is stimulated by estrogen and progesterone. Alveolar cells are columnar and very active in synthesis and secretion. **Myoepithelial cells** surround the alveoli and participate in expulsion of the secretory product. This is stimulated by **prolactin**, **human placental lactogen** and **oxytocin**.

a) Secretory product = milk = proteins + lipids. Lipid material is released by apocrine secretion (part of the apical cytoplasm of the gland cell is lost) and protein material by eccrine secretion (only the secretory granule is released). Proteins are casein, lactoalbumin and immunoglobulins (help protect the feeding infant from GI infections).

MALE REPRODUCTIVE SYSTEM

I. General function. Perpetuation of the species, i.e.,. production of germ cells (sperm)(stimulated by FSH and testosterone) and transport of the sperm to the female for fertilization of the ovum.

II. Components.

A. Primary sex organs. **Testis**. Have dual functions; exocrine (produce sperm) and endocrine (secrete the hormones testosterone, inhibin and androgen binding protein).

B. Secondary sex organs. Epididymis, vas deferens, urethra and penis, seminal vesicles, prostate and bulbourethal glands.

III. Scrotal contents.

A. Testis. Develops retroperitoneally in abdominal cavity, descends to scrotum carrying some peritoneum with it (**tunica vaginalis**). Is suspended by the **spermatic cord** (which contains the vas deferens, cremaster muscle, blood and lymph vessels and nerves).

1. Tunica vaginalis. Covers the anterior two thirds of testis. Composed of 2 layers, visceral and parietal, both of which consist of mesothelium.

B. **Tunica albuginea**. Dense connective tissue layer (capsule) surrounding the testis.

C. **Dartos muscle**. Smooth muscle in the dermis of the scrotal skin. Contraction of the dartos muscle wrinkles the scrotum and contraction of the **cremaster muscle** elevates the testis nearer the body wall; both help maintain temperature for germ cells.

IV. **Testis.**

A. Structure. Tunica albuginea thickens on the posterior aspect of the testis forming the **mediastinum**. From here connective tissue septa radiate into the testis dividing it into lobules; each lobule contains several highly coiled **seminiferous tubules.**

B. Tubules for sperm transport. At the apex of each lobule the seminiferous tubule straightens out as the tubuli recti (straight tubules) which continue in the mediastinum as the rete testis (network of tubules). From the rete testis arise 10-15 efferent ductules which exit the testis and converge to form the ductus epididymis. The ductus epididymis is continuous with the vas deferens (exits the scrotum) which joins with the duct of the seminal vesicle to form the ejaculatory duct. The latter enters the prostatic urethra which continues as the membranous and then penile urethra.

C. **Seminiferous tubules**. Exocrine testis. Consist of a stratified epithelium containing two cell types: germ cells and supporting cells (Sertoli cells). The tubule is surrounded by loose connective tissue containing myoid cells (these contract to help move sperm out of tubules) and small groups of endocrine cells (interstitial cells of Leydig—produce testosterone).
   1. Germ cells. 4-8 layers of cells in various stages of development.
      a) **Spermatogonia**. Closest to basement membrane. Small (12μm) with spherical, dark staining nucleus. Divide by mitosis.
      b) **Primary spermatocytes**. Derived from spermatogonia. Are luminal to the basement membrane, large (18μm) and chromosomes are evident in the nucleus. Undergo the first meiotic division forming 2 secondary spermatocytes.
      c) **Secondary spermatocytes**. Occupy the middle of the epithelium, smaller (12μm), and chromatin is evident. Once formed they divide immediately (second meiotic division) forming 2 spermatids. One spermatogonium forms 4 spermatids.
      d) **Spermatids**. Border the lumen of the tubule. Are small cells (9μm) that do not undergo any further divisions, but undergo maturation (**spermiogenesis**) to become spermatozoa.
   2. **Supporting cells = Sertoli cells.** Tall columnar cells that extend from the basement membrane to the lumen. Have a euchromatic nucleus with prominent nucleoli. Function: have a supportive (trophic) role for germ cells and are secretory.
      a) Secrete **androgen binding protein** (binds high concentrations of testosterone in tubules to promote spermatogenesis) and **inhibin** (feedback inhibition on production of FSH).
      b) Secrete fluid to promote sperm movement out of tubules.
      c) Phagocytose excess cytoplasm shed during spermiogenesis.
      d) Have a nutritive and protective role for developing sperm.
      e) Form the **blood-testis barrier**. Barrier consists of tight junctions between Sertoli cells. Creates a basal compartment (containing spermatogonia) and an adluminal compartment (containing spermatocytes spermatids and spermatozoa). Restricts undesirable macromolecules from reaching spermatocytes and keeps new (antigenic) molecules of spermatocytes and spermatozoa from entering the host immune system and mounting an immune reaction.

D. **Spermatogenesis**. Process of development of spermatogonia to mature sperm. Stimulated by testosterone and FSH. Vitamin A promotes this as does an ideal temperature (35° C.,

214

about 2° C. below body temperature). Takes about 64 days to produce spermatozoa from spermatogonia. This occurs in waves in the seminiferous tubules so that all tubules, and all parts of the same tubule, **are not** in the same stages of spermatogenesis.

E.   **Spermiogenesis**. Maturational process undergone by spermatids to become spermatozoa. Four phases are involved in spermiogenesis and include development of the acrosome (containing lysosomal enzymes necessary for penetration of the ovum), condensation of the nuclear chromatin, development of an elongated cell shape, the flagellum and energy machinery (mitochondria) for motility.

F.   **Mature spermatozoan**. 3 parts: head, neck and tail (flagellum).
1.   **Head**. Condensed nucleus with haploid amount of genetic material. Nucleus is capped by the acrosome containing hydrolytic lysosomal enzymes essential for dispersing the cells of the corona radiata and digesting the zona pellucida of the ovum to aid fertilization.
2.   **Neck**. Contains a centriole and longitudinal fibers continuous into the tail.
3.   **Tail** (flagellum). Comprise 3 parts: midpiece, principal piece and endpiece. Tail is an axoneme (contains the 9 + 2 arrangement of microtubules characteristic of motile cilia). In addition, midpiece contains mitochondria arranged around a system of fibers. Fibers continue into the principal piece.
4.   Spermatozoa undergo a **maturation** in the epididymis and **capacitation** (acquire capacity to fertilize) in the oviduct. Spermatozoa are not mobile until they are exposed to secretions of genital glands (movement through ducts is by fluid currents and peristalsis).

G.   Semen. Seminal plasma (glandular secretions) + spermatozoa.
1.   This is about 3-6 ml of semen per ejaculate.
2.   There is about 100-300 million sperm per ml of ejaculate.
3.   **NOTE**: less than 20 million sperm per ml of ejaculate is considered sterile.

V.   **Tubules.**

A.   **Straight tubules** (tubuli recti). Lined by simple cuboidal epithelium.

B.   **Rete testis**. Network of tubules in the mediastinum. Lined by simple cuboidal epithelium.

C.   **Efferent ductules**. 10-15 tubules exiting the testis. Lined by pseudostratified epithelium characterized by alternating tall columnar (ciliated) and short cells (gives tubule an irregular outline). Smooth muscle among tubules. Sperm movement by ciliary action, fluid movement, and smooth muscle contraction.

D.   **Ductus epididymis**. Formed by convergence of efferent ductules. Lined by pseudostratified epithelium characterized by tall columnar cells (with stereocilia) and basal cells. Columnar cells are secretory. Luminal lining is smooth and regular. Smooth muscle surrounds the duct. Stores sperm, participates in their maturation and propels them into the vas deferens.

E.   **Vas deferens** (ductus deferens). Lined by pseudostratified epithelium like epididymis. Muscularis - 3 well-developed layers of smooth muscle that are highly innervated by sympathetic nerves (undergo powerful contractions for ejaculation). Surrounded by an adventitia (fibrous connective tissue).

VI.   **Glands**.

A.   **Prostate**.

1. Consists of 3 sets of glands organized around the urethra.
   a) **Mucosal**. Contained in mucosa and secrete directly into urethra. With age, these can enlarge (benign), impinge on urethra and impede urination.
   b) **Submucosal**. Extend out from the urethra, but open by ducts into the urethra.
   c) **Main**. Conglomerate of 30-50 tubuloalveolar glands that open by excretory ducts into the prostatic urethra. These produce most of the secretory material of the gland. **NOTE**: main glands are most often involved in malignant carcinoma of the prostate.
2. Epithelium. Low pseudostratified columnar (mucosal and submucosal glands) to simple cuboidal/columnar (main glands). Secretion is part of semen (20-30%), is thin and watery and contains acid phosphatase (elevated in the blood of men with prostatic carcinoma; useful for clinical diagnosis and for monitoring carcinoma), citric acid, amylase, and fibrinolysin (keeps semen liquified).
3. Stroma. Fibroelastic tissue containing smooth muscle. Muscle contractions help expel secretory material.
4. **Prostatic concretions (corpora amylacea)**. Condensations of secretory material often present in the gland lumen. Increase with age.
5. Prostate is dependent on and responds to testosterone.

B. **Seminal vesicle**.
   1. Consists of paired and coiled glandular sacs.
      a) Epithelium is pseudostratified columnar to simple columnar. Mucosa is highly folded and invaginated. Is a secretory epithelium and produces part of semen (70%). Secretion is thick and contains fructose (energy source for sperm), citrate, prostaglandins and flavins (pigments that fluoresce under UV light; can be used in forensic medicine to detect semen stains in suspected rape).
      b) Stroma. Contains many elastic fibers. Smooth muscle is prominent. Contraction helps expel secretory material.
   2. Vas deferens joins the ducts of the seminal vesicles to form the ejaculatory ducts which open into the prostatic urethra.
   3. Seminal vesicle is dependent on and responds to testosterone.

C. **Bulbourethral (Cowper's) glands**. Small glands open in the proximal part of the penile urethra. They secrete a clear, viscous material during sexual arousal.

VII. **Penis**.

A. A. Consists of 3 cylindrical bodies of cavernous or erectile tissue. Cavernous tissue is comprised of a system of endothelial-lined vascular channels. These are engorged with blood during sexual arousal producing penile erection.
   1. **Corpora cavernosa**. Paired and surrounded by a dense connective tissue coat = **tunica albuginea**. Supplied by **deep artery** that give rise to coiled **helicine arteries**. During sexual stimulation sympathetic nerve activity is decreased and helicine arteries dilate, blood pressure overcomes elastic tissue, thus allowing engorgement of cavernous tissue with blood. Inflow exceeds outflow and erection is achieved. After ejaculation, arterial smooth muscle regains its normal tone and blood is dispersed and the penis becomes flaccid.
   2. **Corpus spongiosum**. Single cavernous body that contains the penile urethra. Is surrounded by the tunica albuginea and ends in an expansion, the glans penis. Penile urethra (see urinary system for description).

VIII. Endocrine relationships.

    A.   **Interstitial cells of Leydig**. Have the structure of steroid-hormone producing cells—abundant sER, mitochondria with tubular cristae, lipid droplets and lipofuscin pigment in the cytoplasm.

        1.   Secretion. **Testosterone**. Responds to LH secretion by anterior pituitary gland.

        2.   **Testosterone function**. Stimulates and maintains the structure and function of the secondary sex organs, secondary sex characteristics (deep voice, male hair pattern, and male type chest, pelvis and fat distribution), and spermatogenesis.

    B.   LH (anterior pituitary). Stimulates Leydig cells to produce testosterone. Testosterone has a negative feedback for LH and FSH production.

    C.   FSH (anterior pituitary). Stimulates seminiferous tubules and spermatogenesis directly. Stimulates Sertoli cells to produce androgen binding protein (ABP) and inhibin. Inhibin has a negative feedback to production of FSH. ABP binds and concentrates testosterone in the seminiferous tubules to promote spermatogenesis.

Review Questions

1. Which of the following spinal nerves contain both sympathetic and parasympathetic fibers?

   A. T_{1-12}.
   B. C_{1-8}.
   C. T_1-L_2.
   D. S_{1-5}.
   E. S_{2-4}.

2. Blood flows through the primitive beating heart in the following pattern:

   A. atrium, ventricle, sinus venosus, bulbus cordis, truncus arteriosus.
   B. truncus arteriosus, bulbus cordis, ventricle, atrium, sinus venosus.
   C. sinus venosus, ventricle, atrium, bulbus cordis, truncus arteriosus.
   D. atrium, ventricle, sinus venosus, bulbus cordis, truncus arteriosus.
   E. sinus venosus, atrium, ventricle, bulbus cordis, truncus arteriosus.

3. Lack of development of the thymus or its malfunction during fetal and/or neonatal life will result in:

   A. A decreased number of circulating lymphocytes.
   B. A loss of the graft-rejection phenomenon.
   C. Poor antibody formation in response to antigen.
   D. A and B only.
   E. A, B and C are all correct.

4. Aganglionic megacolon (Hirschsprung's disease) is a congenital disease resulting from the improper development of a portion of the autonomic nervous system. This disease is characterized by a dysfunctional lower colon which results in an inability to evacuate the bowels (defecate). Which neural population would be involved in this disease?

   A. Postganglionic neurons associated with the sacral portion of the parasympathetic division of the autonomic nervous system.
   B. Parasympathetic postganglionic neurons associated with cranial nerve (CN) X.
   C. Sympathetic postganglionic neurons located in the inferior mesenteric ganglion.
   D. Sympathetic preganglionic neurons located in the intermediolateral cell column (lateral horn).
   E. Sensory neurons associated with the sympathetic division of the autonomic nervous system.

5. Vagotomy (resection of the vagus nerve) will result in which one of the following conditions?

   A. The muscularis externa of the stomach will contract more actively than normal.
   B. Pepsin secretion will increase.
   C. HCL secretion will decrease.
   D. The pyloric sphincter will contract.
   E. The salivary glands will produce an excessive and watery secretion.

6. Inability to extend the forearm might indicate a lesion of the:

   A. Ulnar nerve.
   B. Median nerve.
   C. Axillary nerve.
   D. Lateral cord of brachial plexus.
   E. Posterior cord of brachial plexus.

7. Which disease or lesion would NOT likely result in primary muscle atrophy?

   A. Syringomyelia.
   B. Amyotrophic lateral sclerosis.
   C. Transection of the L5 ventral root.
   D. Transection of the L5 spinal nerve.
   E. Tabes dorsalis.

8. Which of the following is more likely to produce lower limb paralysis and loss of pelvic visceral autonomic control?

   A. spinal bifida with meningocele
   B. spinal bifida occulta
   C. tethered spinal cord
   D. spinal bifida with meningomyelocele
   E. dermal sinus

9. All of the following are transmitted within the carpal tunnel EXCEPT:

   A. Flexor pollicis longus tendon.
   B. Median nerve.
   C. Flexor digitorum profundus tendons.
   D. Flexor digitorum superficialis tendons.
   E. Deep branch of ulnar nerve.

10. If the internal carotid artery was occluded, which branch of the external carotid artery would be least effective in providing collateral circulation to the brain?

    A. Facial.
    B. Superficial temporal.
    C. Maxillary.
    D. Ascending pharyngeal.

11. A patient reports to you that he is experiencing generalized muscle weakness which seems to be progressively worsening. With your present knowledge of skeletal muscle morphology and function, indicate which of the following area(s) may be causing the reported problem:

    A. Insufficient norepinephrine is released into the synaptic cleft of the neuromuscular junction.
    B. T-tubules lack an external lamina and do not transmit the neural impulse.
    C. Acetylcholine receptors in the region of the motor end-plate do not function normally.
    D. Muscle spindles are degenerating.
    E. The depolarization signal is not transmitted at the level of the muscle fiber triad.

12. The apex of the heart is located on the anterior chest wall at which intercostal space:

    A. 4th intercostal space.
    B. 5th intercostal space.
    C. 6th intercostal space.
    D. 7th intercostal space.
    E. 8th intercostal space.

13. An infant who is hypocalcemic because of a failure in the embryogenesis of the parathyroid glands also commonly shows faulty development of the:

    A. thyroid gland
    B. submandibular glands
    C. anterior pituitary gland
    D. thymus
    E. posterior pituitary gland

14. Which of the following statements about the external urethral sphincter is TRUE?

    A. It is subject to involuntary control.
    B. It is supplied by parasympathetic fibers arising from pelvic splanchnic nerves.
    C. It is innervated by the pudendal nerve.
    D. It is a modified portion of the pelvic diaphragm.
    E. It surrounds the prostatic portion of the urethra in the male.

15. You have been asked to consult on a patient suspected of having a cerebrovascular accident (CVA, stroke). When dictating the results of your neurological examination, you state that the patient had a loss of pain and temperature sensation on the right side of his face and body, loss of the gag reflex, a positive Romberg's sign and signs of Horner's syndrome on the right side. Your impression from these symptoms is that the patient has experienced a CVA involving the

    _____.

    A. Anterior spinal artery.
    B. Labyrinthine artery.
    C. Posterior cerebral artery.
    D. Posterior inferior cerebellar artery.
    E. Superior cerebellar artery.

16. Surgical treatment for carcinoma of the rectum often requires rectal excision and creation of an abdominal colostomy, utilizing the descending colon. The associated pelvic dissection, which usually ablates the pelvic plexus, commonly results in all of the following deficits EXCEPT:

    A. Decreased sensation in external genitalia.
    B. Inability to ejaculate.
    C. Inability to achieve erection.
    D. Urinary retention and reduction in frequency of voiding.
    E. Decreased peristalsis in the descending colon.

17. In a congenital diaphragmatic hernia all of the following statements are **TRUE EXCEPT**:

    A. The lungs may be hypoplastic secondary to herniation of abdominal viscera into the thorax.
    B. The pleuroperitoneal membranes fail to close the dorsolateral portion of the diaphragm.
    C. The defect is 5 times more likely to occur on the left side.
    D. The developing lung herniates into the abdominal cavity.
    E. A prenatal diagnosis of congenital diaphragmatic hernia can usually be made by sonography.

18. In the process of inflammation in response to an antigen, the role of the neutrophil is:

   A. Release of histamine into the connective tissue in the area of the antigen.
   B. Phagocytosis and neutralization of the antigen.
   C. "Processing" the antigen so that "T" lymphocytes will recognize it as foreign ("non-self").
   D. Initiate the healing process by forming collagen.
   E. Carrying information regarding the antigen to the lymph nodes via lymphatic vessels.

19. Choose the **INCORRECT** statement concerning the heart.

   A. Structurally, the myocardium is comparable to the tunica media of other vessels.
   B. The atrial and ventricular musculature is almost completely separated by the cardiac skeleton.
   C. Purkinje fibers are characterized by the absence of intercalated disks and myofibrils.
   D. Purkinje fibers are located in the subendocardium.
   E. The cardiac skeleton serves as origin and insertion sites for cardiac muscle fibers.

20. All oogonia undergo their last **MITOTIC** division during:

   A. fetal life.
   B. infancy.
   C. the early stages of sexual maturation.
   D. ovulation.
   E. fertilization.

21. A patient presents to the emergency room complaining of a sudden onset of dizziness and headache. During the neurological exam you observe that the patient cannot abduct his left eye and he does not blink when the left cornea is touched with a cotton swab. A lesion in which brainstem location best describes the patients symptoms?

   A. Basis pontis on the left side.
   B. Pontine tegmentum on the left side.
   C. Tegmentum of the mesencephalon on the left side.
   D. Inferior to the pyramidal decussation on the left side.
   E. Inferior to the sensory decussation on the left side.

22. A lesion of the _____ would NOT result in a loss of proprioception involving the right arm?

   A. Left ventroposterolateral nucleus of the thalamus.
   B. Right medial lemniscus.
   C. Right fasciculus cuneatus.
   D. Left postcentral gyrus of the cerebral cortex.
   E. Right C8 to T1 dorsal root ganglia.

23. In the developing neural tube, neuronal differentiation occurs primarily in the:

   A. mantle layer.
   B. marginal layer.
   C. ependymal layer.
   D. ependymal, mantle and marginal layers.
   E. ependymal and marginal layers.

24. Which of the following statements describes events which occur **SIMULTANEOUSLY** during the menstrual cycle?

    A. Follicles develop, endometrium proliferates and the vaginal epithelium thickens.
    B. Follicles develop, endometrium becomes secretory and the vaginal epithelium thickens.
    C. Follicles develop the theca interna, progesterone secretion begins and the cervical mucosa is shed.
    D. Follicles develop, the endometrium involutes, and the vaginal luminal pH increases.
    E. Follicles develop, corpora albicans are activated, estrogen and progesterone are secreted.

25. Which statement does NOT characterize a neuron undergoing a reaction to injury?

    A. A centrally located nucleus.
    B. Wallerian degeneration.
    C. A swollen perikaryon.
    D. A reduction in Nissl staining.
    E. Chromatolysis.

26. Case History: A couple in their mid-twenties consult you because the woman has not been able to become pregnant. Tests show that the man is potent and fertile. Precise information regarding menstrual regularity is not available. A biopsy of the ovary shows many atretic follicles, some developing follicles but no corpora lutea or corpora albicans.

Which condition below would provide an explanation for the biopsy findings?

    A. No theca interna to secrete estrogen.
    B. No LH surge at midcycle.
    C. No FSH secreted by the pituitary during the proliferative stage.
    D. No prolactin secreted by the pituitary during the secretory stage.
    E. The relative lack of progesterone during the secretory phase prevents the completion of meiosis.

27. Which of the following organs is NOT a part of the sympathetic division of the autonomic nervous system?

    A. Aorticorenal ganglion.
    B. Celiac ganglion.
    C. Suprarenal medulla.
    D. Inferior mesenteric ganglion.
    E. Myenteric plexus ganglia of the stomach.

28. A patient has pernicious anemia resulting from an inadequate absorption of vitamin B12. Which of the following cells would most likely be malfunctioning?

    A. Paneth cells of the small intestine.
    B. Enteroendocrine cells of the stomach.
    C. Cells lining the intestinal glands.
    D. Parietal cells of the gastric mucosa.
    E. Chief cells of the gastric mucosa.

29. The formation of platelets involves which of the following?

    A. Loss of the megakaryocyte nucleus.
    B. Proliferation of platelets in the marrow stroma.
    C. Erythropoietin.
    D. Fragmentation of megakaryocyte cytoplasm.
    E. Secretion of thromboplastin.

30. General somatic afferent fibers are found in all of the following EXCEPT:

    A. All dorsal roots.
    B. All cutaneous nerves.
    C. All muscular nerves.
    D. All dorsal root ganglia.
    E. All sympathetic splanchnic nerves.

31. The normal abdominal body wall of the adult is formed by contributions from:

    A. ectoderm and lateral plate somatic mesoderm.
    B. ectoderm and lateral plate splanchnic mesoderm.
    C. ectoderm, lateral plate somatic mesoderm and paraxial mesoderm.
    D. ectoderm, lateral plate splanchnic mesoderm and paraxial mesoderm.
    E. ectoderm and neural crest cells.

32. Which statement is NOT correct?

    A. Occluding the left anterior cerebral artery may result in anesthesia the ipsilateral foot.
    B. The primary visual cortex is supplied, in part, by the posterior cerebral artery.
    C. Occluding the right middle cerebral artery may result in a loss of motor function in the contralateral forearm.
    D. The ophthalmic artery is a direct branch of the internal carotid artery.
    E. The inferior sagittal sinus does not drain directly into the confluence of sinuses.

33. Which of the following correctly characterize B-lymphocytes?

    A. Can differentiate into plasma cells.
    B. Migrate readily and freely in the blood vascular and lymph vascular systems.
    C. Form about ninety percent of the lymphocyte population.
    D. Form the predominant population of lymphocytes in the thymus gland.
    E. All of the above are correct.

34. Structures found in the lateral wall of the cavernous sinus include all of the following EXCEPT:

    A. Oculomotor nerve.
    B. Trochlear nerve.
    C. Ophthalmic division of trigeminal nerve.
    D. Maxillary division of trigeminal nerve.
    E. Mandibular division of trigeminal nerve.

35. Radiation is means of inhibiting the proliferation of cells. Which of the following cell populations most likely would be reduced in a 10-year-old patient receiving radiation therapy of a tumor in the bone marrow cavity?

   A. Osteocyte population.
   B. Adipocyte population.
   C. Marrow blood vessel endothelial cell population.
   D. Osteoblast population.
   E. Skeletal muscle cell population in the path of the radiation.

36. Which of the following statements **CORRECTLY** pertain to alveolar pores:

   A. Function in collateral air circulation.
   B. Are part of the gas exchange mechanism.
   C. Help to increase the blood perfusion of the lung.
   D. Serve to equalize the interalveolar air pressure.
   E. A and D are correct; B and C are incorrect.

37. The pectinate line of the anal canal is the line of junction or separation between all of the following EXCEPT:

   A. Venous drainage to portal and caval systems.
   B. Somatic afferent and visceral afferent fibers.
   C. Middle rectal arterial supply and inferior rectal arterial supply.
   D. Derivatives of ectoderm and endoderm.
   E. Lymphatic drainage to inguinal nodes and inferior mesenteric nodes.

38. Tetralogy of Fallot includes all of the following except:

   A. pulmonary stenosis.
   B. large foramen ovale.
   C. right ventricular hypertrophy.
   D. dextropositional (overriding) aorta.
   E. large ventricular septal defect.

39. If the dermis of the integument is traumatized (cut in the skin) which of the changes listed below would you NOT expect to happen during the ensuing days?

   A. Fibroblasts would increase in number.
   B. Macrophages would be engaged in phagocytosis.
   C. Fibroblasts would become more basophilic.
   D. Mast cells would appear in large numbers and would be releasing their granules.
   E. New collagen fibers will be synthesized.

40. The anastomosis between the superior and inferior mesenteric arteries is near the:

   A. Sigmoid colon.
   B. Descending colon.
   C. Hepatic flexure.
   D. Rectum.
   E. Splenic flexure.

41. Which of the following structures do not form part of the filtration membrane in the renal glomerulus?

   A. Capillary endothelial cell.
   B. Parietal layer of Bowman's capsule.
   C. A basement membrane (a basal lamina).
   D. Podocytes.
   E. Filtration slits.

42. Following partial resection of the thyroid gland, a patient had paralysis of the cricothyroid muscle. This was most likely due to:

   A. Severing the inferior laryngeal nerve.
   B. Severing the vagus nerve.
   C. Severing the superior laryngeal nerve.
   D. Trauma to the muscle during surgery.
   E. Secondary infection subsequent to surgery.

43. Which cell does NOT have a developmental origin from ectoderm?

   A. Chromaffin cells of the adrenal medulla.
   B. Microglia.
   C. Oligodendroglial cells.
   D. Purkinje neurons of the cerebellum.
   E. Ependymal cells lining the central canal of the spinal cord.

44. Failure of the embryonic mesoderm to invade and fuse across the midline cranial to the oropharyngeal membrane (prochordal plate) would be expected to interfere with normal development of all of the following **EXCEPT**:

   A. heart.
   B. pericardial cavity.
   C. septum transversum.
   D. neural tube.
   E. thoracic body wall.

45. Certain drugs, such as colchicine or vinblastine, will depolymerize or impair the function of microtubules. Based upon this information and your knowledge of the function of microtubules, ALL of the following statements would correctly pertain to the effects of having administered one of these drugs to an individual in cancer chemotherapy **EXCEPT**:

   A. Lack of ciliary motility.
   B. Impaired mitosis.
   C. Impaired axonal transport in neurons.
   D. Lack of movement of mRNA into the cytoplasm.
   E. Impaired secretion of proteins packaged for export.

46. Gastrulation in the human embryo:

    A. is the process by which the bilaminar embryo becomes trilaminar.
    B. involves the movement of epiblast through the primitive streak.
    C. gives rise to all embryonic endoderm.
    D. gives rise to all embryonic mesoderm.
    E. all of the above are correct.

47. In precocious sexual development, skeletal maturation and growth is stunted primarily due to:

    A. Delay of epiphyseal union.
    B. Decrease in sex hormone secretion.
    C. Increase in growth hormone secretion.
    D. Premature epiphyseal union.
    E. Deficiency of vitamin C.

48. Concerning the latissimus dorsi muscle, all of the following are true EXCEPT:

    A. Acts in extension of the humerus.
    B. Inserts in the intertubercular groove of the humerus.
    C. Helps to form the posterior axillary fold.
    D. Is innervated by a nerve from the posterior cord of the brachial plexus.
    E. Acts in lateral rotation of the humerus.

49. All of the items in the column on the right pertain best to which **ONE** item in the column on the left?

    A. Zona fasciculata
    B. Parathyroid gland
    C. Oxyphil cells
    D. Adrenal medulla
    E. Paraventricular nucleus

    1. Cells are of neural crest origin
    2. Produce epinephrine
    3. Contains a central vein
    4. Shows a positive chromaffin reaction

50. You observe urine leaking out of the umbilicus of a neonate. Which of following malformations could account for this observation?

    A. a patent urachus
    B. a patent vitelline duct
    C. lack of closure of the medial umbilical ligaments
    D. exstrophy of the bladder
    E. an omphalocele

51. All of the items in the column on the right pertain best to which **ONE** item in the column on the left?

    A. Epididymis
    B. Prostate gland
    C. Corpus cavernosum
    D. Seminal vesicle
    E. Rete testis

    1. Simple columnar epithelium.
    2. Morphology may be stimulated by testosterone.
    3. Produces part of the semen.
    4. Stroma contains smooth muscle.
    5. Produces significant amounts of acid phosphatase.

52. A patient arrives at your clinic with signs and symptoms of Parkinson's disease. Which statement is NOT appropriate for this disease?

    A. The patient's arm moves in a jerky motion when flexed against resistance.
    B. At rest, you notice that the patient has a pronounced tremor involving his fingers.
    C. The patient is taking dopamine in order to lessen his symptoms.
    D. There is a reduction in the number of dopamine-containing neurons in the pars compacta of the substantia nigra.
    E. The patient has a shuffling gait with an apparent decrease in facial expression.

53. Vitamin C is important for normal growth of cartilage because it:

    A. Stimulates normal proliferation of chondroblasts.
    B. Is required for absorption of calcium by the intestine.
    C. Is necessary for sulfate endocytosis.
    D. Is necessary for maintenance and synthesis of collagenous fibers.
    E. Is needed for polymerization of chondroitin.

54. First arch syndrome is believed to result from insufficient migration of neural crest cells into the first branchial arch. All of the following malformations could result from this syndrome **EXCEPT**:

    A. hearing loss due to a malformed malleus.
    B. a cleft palate.
    C. low set ears.
    D. lack of a thymus.
    E. a small mandible.

55. Which statement best characterizes the neurological symptom typical of a patient with an enlargement of the pituitary gland (due to an adenoma)?

    A. An inability to rotate the eyes laterally.
    B. An interruption of neural signals from the temporal retina.
    C. A loss of the accommodation/convergence reflex.
    D. Night blindness and a generalized decrease in the perception of colors.
    E. A decrease in peripheral vision.

56. Violent involuntary, flailing movements involving the proximal extremities and limb girdles characterize which syndrome, disease or specific lesion?

    A. Huntington's chorea.
    B Multiple sclerosis.
    C. Sydenham's chorea.
    D. Tic douloureux.
    E. Hemiballism.

57. Which of the following statements about nondisjunction is **INCORRECT**?

    A. Nondisjunction that occurs during mitotic cell division results in a mosaic.
    B. Nondisjunction may involve either autosomes or sex chromosomes.
    C. Both monosomic and trisomic conditions may result from nondisjunction.
    D. Nondisjunction resulting in trisomy of autosomes is invariably lethal.
    E. Nondisjunction in germ cells produces at least two chromosomally abnormal gametes.

58. When viewing your patient's magnetic resonance image (MRI), you notice a lesion in her medulla oblongata on the left side. Given that the lesion impairs cranial nerve function, which statement describes ani nappropriate action to the test performed?

    A. When asked to say, "Ahh", her uvula deviates to the right.
    B. When protruding the tongue, it deviates to the left.
    C. When asked to shrug her shoulders against resistance, there is a significant reduction in muscle strength on the left side.
    D. There is a significant decrease in her ability to turn her head to the left against resistance.
    E. She experiences pain, but does not blink when her left cornea is touched with a cotton swab.

59. Which statement is NOT correct?

    A. The dendritic trees of cerebellar Purkinje neurons are oriented perpendicular to the long axis of the folia.
    B. Only a minority of corticospinal tract axons originate from neurons located in the precentral gyrus.
    C. Lesions of Broca's area on the dominant side result in receptive, but not expressive aphasia.
    D. Herniation of the cerebellar tonsils into the foramen magnum is indicative of an increase in intracranial pressure.
    E. Climbing fibers in the cerebellum originate from neurons located in the inferior olivary nucleus.

60. A high school baseball player sustained an injury sliding into second base. As a sequelae to this injury he was unable to dorsiflex or evert his left foot. Cutaneous sensation over the dorsum of the left foot and anterolateral aspect of the lower half of the left leg was impaired. The injury most likely involved the:

    A. Common peroneal nerve.
    B. Tibial nerve.
    C. Femoral nerve.
    D. Obturator nerve.
    E. Inferior gluteal nerve.

61. The sclerotomic portions of somites make contributions to all of the following **EXCEPT** the:

    A. vertebrae.
    B. sacrum.
    C. occipital bone.
    D. long bones of the upper extremity.
    E. secondary sclerotome.

62. Following an injury to the spinal cord, a patient experienced a complete loss of touch and vibration sense and motor function in his right leg. However, he was able to feel pain when areas of his right leg were pricked by a pin. Motor function and touch sensation were intact in the left leg;, but a significant decrease in the patient's sensitivity to painful stimuli was observed. Select the best answer that would explain the patient's symptoms?

    A. A complete transection at the T6 spinal cord segmental level.
    B. A complete transection at the L5 spinal cord level.
    C. A hemisection of the left spinal cord at the S2 spinal cord segmental level.
    D. A hemisection of the right spinal cord at the L1 spinal cord segmental level.
    E. A hemisection of the right spinal cord at the S1 spinal cord segmental level.

228

63. A patient has a right heart failure with retrograde impaired drainage of the inferior vena cava and hepatic veins. Which part(s) of the structural liver lobule would you expect to undergo primary atrophy first?

A. Peripheral part.
B. Area around the bile ducts.
C. Area around the portal triangle.
D. Central part.
E. Area around the terminal branches of the hepatic artery and portal vein.

64. A lesion of the middle cerebral peduncle would interrupt which tract or projection?

A. Ventral spinocerebellar tract.
B. Projections from the cerebellum to the inferior olivary nucleus.
C. Cortico-ponto-cerebellar fibers.
D. Trigeminocerebellar tract.
E. Projections from the cerebellum to the red nucleus.

Questions 65-70: Use the following list to answer these questions.

(A)   Congenital aganglionic megacolon
        (Hirschsprung disease)

(B)   Duodenal atresia

(C)   Congenital pyloric stenosis

(D)   Thyroglossal duct cyst

(E)   Tracheoesophageal fistula

(F)   Bilateral renal agenesis

(G)   Spinal bifida cystica

(H)   Urachal cyst

(I)   Meckel's diverticulum

(J)   Congenital umbilical hernia

(K)   Umbilico-ileal fistula

(L)   Spinal bifida occulta

(M)   Branchial cyst

(N)   Congenital omphalocele

For each patient described, select the appropriate congenital malformation.

65. An infant begins vomiting a few hours after birth. The vomitus contains bile. There is accompanying distention of the epigastrium.

66. An amniocentesis is preformed fifteen weeks after the last normal menstrual period. There is a high level of alpha-fetoprotein found in the amniotic fluid.

67. A one week old infant begins to develop a mass that protrudes through the umbilicus during crying, straining or coughing.

68. A low volume of amniotic fluid, oligohydramnios, is observed during the third trimester.

69. A three year old boy begins to develop a painless, progressively enlarging, and movable mass in the median plane of the neck just inferior to the thyroid gland.

70. A one day old infant has abdominal distension after crying, respiratory distress, and total inability to feed without coughing and regurgitation.

Answers to Review Questions

1.	E	25.	A	49.	D
2.	E	26.	B	50.	A
3.	E	27.	E	51.	B
4.	A	28.	D	52.	C
5.	C	29.	D	53.	D
6.	E	30.	E	54.	D
7.	E	31.	C	55.	E
8.	D	32.	A	56.	E
9.	E	33.	A	57.	D
10.	B	34.	E	58.	D
11.	C	35.	D	59.	C
12.	B	36.	E	60.	A
13.	D	37.	C	61.	D
14.	C	38.	B	62.	D
15.	D	39.	D	63.	D
16.	A	40.	E	64.	C
17.	D	41.	B	65.	B
18.	B	42.	C	66.	G
19.	C	43.	B	67.	J
20.	A	44.	D	68.	F
21.	B	45.	D	69.	D
22.	B	46.	E	70.	E
23.	A	47.	D		
24.	A	48.	E		